THE TRAGEDY OF EVOLUTION

The Human Animal Confronts Modern Society

Michio Kitahara

New York
Westport, Connecticut
London

Copyright Acknowledgment

Acknowledgment is gratefully made to JAI Press for permission to reprint portions of the following article in Chapters 1, 3, and 4:

Michio Kitahara, "Sociocultural Phenomena as Stimulus Seeking Behavior," copyright 1991 by JAI Press, Inc., first published in the *Journal of Social and Biological Structures*, 14 (1991).

Library of Congress Cataloging-in-Publication Data

Kitahara, Michio.
    The tragedy of evolution : the human animal confronts modern society / Michio Kitahara.
      p. cm.
    Includes bibliographical references and index.
    ISBN 0–275–94041–1 (alk. paper)
    1. Social evolution. 2. Sociobiology. 3. Civilization,
Occidental.   I. Title.
GN360.K5   1991
304.5—dc20   91–10180

British Library Cataloguing in Publication Data is available.

Library of Congress Catalog Card Number: 91–10180
ISBN: 0–275–94041–1

First published in 1991

Praeger Publishers, One Madison Avenue, New York, NY 10010
An imprint of Greenwood Publishing Group, Inc.

Printed in the United States of America

The paper used in this book complies with the
Permanent Paper Standard issued by the National
Information Standards Organization (Z39.48–1984).

10 9 8 7 6 5 4 3 2 1

THE TRAGEDY OF EVOLUTION

To those who cherish individualism
against every form of tyranny

CONTENTS

PREFACE

One's philosophical and ideological outlook is to a significant extent shaped by the intellectual climate in which one lives. It is certainly misleading and incorrect to use such expressions as "conditioned" or "determined" in this context, but I feel that the current intellectual climate of Western civilization often presents a misleading picture of its future. In particular, I would like to point out two beliefs in Western civilization that probably should be carefully scrutinized and reconsidered. These are (1) biological and cultural evolutionism, and (2) the ultimate basis of human behavior.

As a product of the philosophy of the Enlightenment, evolutionism is unduly optimistic, involving such assumptions as improvement, refinement, and progress. Certainly, biological evolutionism has been questioned and criticized more and more in recent years, as summarized, for example, by Francis Hitching in his very informative book *The Neck of the Giraffe*. But my question deals with the optimism of evolutionism *per se*. Cultural evolutionism, too, tends to emphasize only the positive aspects of evolutionism.

The basis of human behavior is another belief in Western civilization that should be carefully reconsidered. Above all, it is believed that we differ significantly from other animal species, and that our genetic background means little in understanding our behavior. We believe our existence is in a dimension totally different from that of other animal species. In this book, I challenge these beliefs of Western civilization and present my own view.

However, my ideas cannot be presented as a book on the basis of my own effort alone. First of all, I would like to express my sincere gratitude to Professor P. J. Vinken, president of Elsevier in Amsterdam, who found some value in my work and called Praeger's attention to my manuscript. Professor Albert Somit, president of Southern Illinois University at Carbondale, was kind enough to offer me encouragement, suggestions, and moral support when I most needed them.

Practically all of my research for the preparation of this book was conducted at the main and specialized libraries of the University of Gothenburg, where I always was offered very competent, efficient, courteous, and friendly assistance. I would like to thank each one of the personnel at these libraries. In the practical matters of editing, production, and other things to be done, I would like to thank Mrs. Anne Kiefer and Mrs. Nina Neimark at Praeger.

I realize that a book cannot be published without the good will of many human beings, and I, as another human being, am happy to be able to thank them.

INTRODUCTION

We are scientifically called *Homo sapiens sapiens*. *Homo sapiens sapiens* is a subspecies of *Homo sapiens*, which includes other subspecies of the past such as *Homo sapiens fossilis* and *Homo sapiens neanderthalensis*. We are the only surviving subspecies of *Homo sapiens*. This may make us feel complacent, but that is quite wrong.

We are faced with problems of all kinds in the world today. There are personal, interpersonal, regional, national, and international problems, such as alienation, parent-child conflict, racial, ethnic, and religious conflict, strikes, unemployment, high crime rates, nuclear accidents, inflation, trade wars, environmental pollution and destruction, and so on. Certainly, no one has any difficulty in making a long list of this nature. These are not problems limited to certain cultures or nations but rather are found practically everywhere in the world.

Of course, I am not the first one to consider them as problems. Already there are many books about our problems, discussing them from diverse perspectives such as philosophy, theology, sociology, economics, anthropology, psychology, psychiatry, ethology, and sociobiology. However, except for certain books written by ethologists and sociobiologists, most of these books do not look at our problems in terms of our biological background. Since we are a biological species, whether we like it or not, we cannot get away from this fact, and ultimately our existence is based on our background as an animal species. I have written this book from such a point of view.

At the same time, we are a species with advanced culture. Many studies now abundantly indicate that we are no longer the only species having culture. But no doubt we have advanced the most in building up culture. Some of our cultural products are purely symbolic, ideological, and philosophical. The idea of freedom and liberty, for example, as discussed by liberals such as J. S. Mill, cannot be found among chimpanzees, and certainly this is quite unique to *Homo sapiens sapiens* among the numerous living species.

The purpose of this book, then, is to look at the complex phenomenon of our culture in terms of our biological background. I would like to examine our problems by constantly referring back to our biological background as an animal species. Methodologically, this approach may sound similar to the ones taken by ethologists or sociobiologists, but I make my own assumptions about human beings as a species, and I do not share most of their views.

The assumptions I make in this book are very simple, and I consider them to be axiomatic. There are two such assumptions. First, I assume that we, as an advanced primate species, are characterized by a form of behavior that some zoologists call "stimulus seeking behavior." This refers to the animal's curiosity and attempt to know its environment, to interact with it, and to try to use it to its own advantage. Since this behavior involves parts of its own body such as the limbs, the nose, the tongue, or the tail, it is logical to assume that stimulus seeking behavior is accompanied by a drive, which is often called "manipulatory drive." The manipulatory drive is not necessarily goal-directed. An animal may satisfy its manipulatory drive by merely releasing it toward the environment. In reality, it is difficult to separate stimulus seeking behavior from the manipulatory drive, and these two are likely to be found together.

Second, I also assume that, as a biological species, we are characterized by variation. Variation is necessary for an organic species, whether it be an animal or a plant, because variation allows the species to adapt to environmental change. Without variation, a species has very little chance for survival. These two assumptions—namely, (1) that humans are characterized by stimulus seeking behavior accompanied by the manipulatory drive, and (2) that humans are characterized by variation—are the key assumptions of this book, as it considers our culture and our problems. I shall briefly summarize how I look at our situation in terms of these two assumptions.

SUMMARY OF THE BOOK

Stimulus seeking behavior includes a variety of forms, such as behavior based on curiosity, as well as play and creativity. Although many species show such behaviors, it is typically found in mammals, and especially in primates. Within primates, there tends to be an association between the extent of stimulus seeking behavior and the level of evolution. It is logical to assume, then, that we have genetically acquired a strong tendency for active stimulus seeking behavior, accompanied by the corresponding drive to manipulate the environment.

We manipulate both material and symbolic objects. Technology is a means to manipulate, and culture can be considered as an established form of stimulus seeking behavior. Aggression is an extreme form of the attempt to manipulate, in which one tries to manipulate objects violently.

The development of the manipulatory drive is ontogenetically seen in the growth of the human child. Conflict between parent and child includes two forms. First, it is an incompatibility between the manipulatory drives of two individuals, the parent and the child, and, second, it is also the conflict between the child's manipulatory drive and the parent's protective behavior derived from the evolution of the phylum Chordata for more efficient and effective reproduction. The psychological and psychoanalytic mechanism of identification evolved to deal with conflict of this nature.

Deviant behavior, such as criminal behavior, sexual deviance, gambling, as well as an abnormal concern with or interest in money, power, the computer, the automobile, and so on, are stimulus seeking behavior in excessive or unusual forms. Such behavior, and especially juvenile delinquency in the form of vandalism and shoplifting, is typical of the stimulus seeking behavior called "play." Both juvenile delinquency and play are behaviors found among youth and are more common among males than among females. So-called white-collar crime involves manipulation of symbolic objects. Alienation occurs when one is unable to express one's manipulatory drive satisfactorily. Some forms of suicide are due to the attempt to manipulate one's own body. This can be inferred from the fact that during a war, which allows people to satisfy their manipulatory drive vicariously, suicide rates drop. Also men, who biologically have a stronger manipulatory drive than women, commit suicide more.

Conflict in society is a result of incompatibility of the manipulatory drives among people. In the West, after the rise of the philosophy of the Enlightenment, more objects were seen to be manipulatable, and people began to attempt active manipulation of their environment. The ideology of democracy further encouraged this trend, and such sociological phenomena as status incongruence and relative deprivation became common, resulting in, for example, revolution.

Some forms of prejudice and discrimination result from conflict over the manipulation of resources, and the manipulatory drive may be directed against substitute objects in the form of scapegoating. Conflict may result from the manipulation of someone else's part of the self, which William James called the material self. Society often punishes a person by intentionally manipulating his or her material self.

Variation among us has three dimensions: physical, psychological, and cultural. Because of these three forms of variation, conflict is natural. Statistically speaking, variation takes the form of the bell-shaped distribution curve, in which the majority of individuals fall in the middle, while a small number of individuals lie on both sides of the curve. This means two things. First, it is possible to recognize the tendency for the majority, and, second, extreme individuals function as "banks."

Human variation is gradual rather than abrupt. For this reason, the categorization of individuals is possible on the basis of their apparent similarities. But we tend to be deceived by this because we ignore differences. Controversies over the publications by James Coleman, Arthur Jensen, and Christopher Jencks are examples. What Daniel Bell calls "equality of outcomes" is another product of the same mistake. Since the time of Locke, Bentham, and J. S. Mill, the ideology of democracy and equality has changed its character because of this false thinking.

Our cultural evolution can be heuristically seen in three phases: (1) hunter-gatherer society, (2) agricultural society, and (3) industrial society. *Homo sapiens sapiens* has been biologically the same for at least 35,000 years, and, during this period, we have had the same potential for psychological variation. But our thinking is to a considerable extent a response to our physical and cultural environments, and, for this reason, psychological variation among hunters and gatherers was limited.

After the neolithic revolution, the physical and cultural variations became greater, and so was psychological variation, to some extent. But it was the industrial revolution that truly activated the potential for psychological variation among us by making physical and cultural variations much greater.

Ultimately, technology has been the key factor in increasing physical, psychological, and cultural variations. But by reflecting our primatological background, technology is applied from the perspective of the individual. Technology is seen only in the limited range of its consequences, or what G. H. Mead calls the "manipulatory area." This is the very reason why our environment is being destroyed. Stimulus seeking behavior and the corresponding manipulatory drive were adaptive for hunting and gathering, but, ironically, this biological heritage is now hurting us, and the condition is getting worse.

Both liberalism and capitalism are the products of the philosophy of the Enlightenment, and in the beginning there was no apparent logical inconsistency between them. But capitalism encouraged the development of more and more powerful and effective technology. Technology is a means of manipulation, and the way it is applied may not be acceptable to everyone. Yet it has come to dominate our lives more and more, and, in this sense, capitalism is crushing liberalism. Capitalism did this also indirectly, by encouraging the rise of socialism and communism, both of which are by nature against liberalism.

Democracy is another product of the philosophy of the Enlightenment, but, through the tyranny of the majority, it is also a threat to liberalism. What Isaiah Berlin calls "negative freedom" is a recent product of Western thought after industrialization. This idea emerged in response to the increased physical, psychological, and cultural variations in industrial society. This is the idea of the absence of coercion by others advocated by classical liberals such as J. S. Mill and Benjamin Constant, but, ironically, it is now being crushed by its intellectual sibling, democracy.

In the past, negative freedom was relevant mostly in connection with politics, but its necessity now is greater than ever because of the threat of technology, such as environmental destruction, the development of high technology, and biotechnology, all of which concern everyone in the world. Technology almost always works by promoting "freedom for" at the expense of "freedom from" (negative freedom). Technology is killing negative freedom, because the idea of negative freedom has never been held by the majority, and the tyranny of the majority kills such an idea.

These evolutionary changes have happened and are happening because of the two biological phenomena we genetically inherited, namely (1) stimulus seeking behavior and the corresponding manipulatory drive, and (2) physical, psychological, and cultural variations. We suffer as a result of evolution instead of enjoying its benefits.

Of course, we suffered in the past, too. But in the past the sources of suffering were mainly in the ecological and geographical sphere, such as the elements and natural catastrophes, which are (1) external to people, (2) objective and material, and (3) applicable to everyone. But the nature of suffering has changed; the sources of suffering are now (1) internal to people, (2) subjective and nonmaterial, and (3) not applicable to everyone. This is exactly the reason why negative freedom is urgently needed now more than ever before in order to reduce this new form of suffering.

Evolution is tragic because (1) the manipulatory drive has created technology that is now too powerful; (2) this drive is incompatible with human variation; (3) we know this incompatibility is a source of suffering;

(4) the incompatibility is becoming more and more intense; and (5) human variation itself has become a source of suffering; and as a result, (6) we are now even trying to eliminate physical, psychological, and cultural variations, and (7) the ideology of liberalism, which is the most realistic ideology in view of human variation, is being killed by its ideological sibling, democracy, through the latter's tyranny of the majority.

Western civilization acquired its uniqueness through the philosophy of the Enlightenment, which produced advanced technology and liberalism. But the logical inconsistency within Western civilization has become the source of its crisis. Because of its overwhelming importance in the world today, the crisis of Western civilization is almost synonymous with the crisis of *Homo sapiens sapiens*.

1

STIMULUS SEEKING BEHAVIOR
AND THE MANIPULATORY DRIVE

One of the common characteristics of life is that a living organism responds to stimuli in the environment. Even a unicellular organism such as a paramecium moves in response to changes in the immediate environment in an un-oriented manner known as "kinesis." By responding in this way, a paramecium increases the possibility for survival by avoiding an unfavorable condition and seeking a more favorable one.

At the level of the unicellular organism, the relationship with the environment is simple and passive. The organism merely seeks or avoids stimuli, and it does not try to change the environment. But, through evolution, organisms have become more active in their relationship with the environment. The sensory organs have become refined and sophisticated, and a clear tendency for active manipulation of the environment has evolved.

STIMULUS SEEKING BEHAVIOR IN MAMMALS

In the study of animal behavior, one type of behavior is known as "stimulus seeking."[1] Stimulus seeking behavior is widely known among all orders of mammals,[2] and it is very comprehensive in meaning, including such behaviors as exploration, play, and creativity. Exploration is a preliminary behavior toward the environment, in which an animal focuses its attention on a stimulus cautiously and tentatively. The animal tries to find out about the stimulus and to obtain more information about it.[3] On the basis of knowledge derived from exploration, the animal may use the stimulus object or interact with it, and such behavior is called play. By means of play, the animal derives more stimulation. In the case of creativity, the animal produces novel effects, which also stimulate the animal.

Probably most of us have seen a cat in the following situation. A cat, especially a kitten, may discover, for example, a tennis ball on the floor as a novel object. First the kitten realizes that such an object exists on the floor and looks at it for a while. Then it approaches the tennis ball carefully, trying to find out if it moves or is dangerous. This is exploration. Assuming that the tennis ball neither moves nor is dangerous, the kitten comes closer and closer and finally touches it.

When the kitten touches the tennis ball, it moves. The kitten becomes surprised by this unexpected outcome and runs away about a foot and a half and looks at it again from a distance. Recognizing that the tennis ball stopped moving, the kitten approaches it again, still cautiously. When the kitten touches the ball, it moves again, but this time the kitten is not surprised. After realizing that the action of touching the ball makes it move, the kitten repeats the same action many times. This simple, spontaneous reaction is play.

By intentionally touching the ball, the kitten sees the ball roll away some distance. By chasing the rolling ball, the kitten derives a new experience; the ball is moving at the same time the kitten is moving, and in the same direction. The kitten's movement is directed toward a moving object. In this way, the kitten has created a novel effect by means of its own action. This is an example of creativity. The kitten consciously and intentionally moves toward the ball rolling away from it, instead of merely having a spontaneous response to the moving ball, as in the case of play. In this creative behavior of this kitten, the tennis ball has become an object to chase, giving the kitten a novel experience.

What is the meaning of stimulus seeking behavior for animals? One way of answering this question is to consider whether or not such behavior is adaptive. If we look at the scientific literature dealing with stimulus seeking behavior, there is reason to assume that animals can survive better through stimulus seeking behavior. By definition, stimulus seeking behavior refers to the form of behavior in which animals try to acquire more knowledge about the environment, to establish a way of dealing with it, and even to manipulate it.

Logically, behavior of this kind contributes to the animal's survival. To know the environment better may mean the discovery of phenomena that can be utilized in order to survive as well as the avoidance of phenomena that are dangerous and harmful for survival. It is quite understandable that those mammals that have evolved the greatest capacity for learning are also the ones that play most often.[4]

THE FUNCTIONS OF STIMULUS SEEKING BEHAVIOR

John D. Baldwin and Janice I. Baldwin have reviewed the scientific literature dealing with exploration and play and have recognized 30 functions of stimulus seeking behavior under four major categories: (1) physical development, (2) psychological benefits, (3) coping with the nonsocial environment, and (4) social adjustment. They consider these 30 functions all to apply to primates. The following is a summary of their stance.[5]

Physical Development. Through stimulus seeking behavior, the animal uses its muscles, and in this way such behavior helps it to develop physically.

Psychological Benefits. Stimulus seeking behavior contributes to the development of the nervous system, to its sensitivity and effectiveness, and to the development of perceptual skills, motor skills, and coordination.

Coping with the Nonsocial Environment. The animal can learn about the environment, discover how the environment responds to its behavior, and develop mastery and competence toward the environment.

Social Adjustment. The animal can become integrated into the group, develop social ties, develop a normal personality, learn that it belongs to a certain species, learn communication skills, develop social perception, learn group traditions, learn adult behavior, develop sex roles and reproductive skills, learn dominant and subordinate relations, and learn to control or release aggression.

Stimulus seeking behavior may at times result in negative consequences, which can work against the survival of individual animals. For example, among olive baboons, juvenile males are reported to suffer the highest mortality as a result of their exploration and play.[6] To explore means to be exposed to a new situation, one unfamiliar to the animal, and this naturally entails the possibility for dangers unknown to the animal. To remain in the safe area excludes the possibility for a new experience, and the advantage for survival resulting from stimulus seeking behavior is also excluded. Therefore, despite the situations in which stimulus seeking behavior may mean danger, we may assume its advantage is greater than its disadvantage for a particular species as a whole.

STIMULUS SEEKING BEHAVIOR AND EVOLUTION

It is known that stimulus seeking behavior is generally associated with level of evolution.[7] The species that are more advanced evolutionarily tend to be more active in stimulus seeking behavior. For example, according to

E. W. Menzel, in a 0.9 acre compound, chimpanzees explored and manipulated a much greater variety of objects, compared with rhesus monkeys in the same environment. The chimpanzees showed much more complex stimulus seeking behavior than the rhesus monkeys, and their way of examining the environment was far more detailed.[8]

A similar observation can be made regarding primates in the natural habitat. It is estimated that savanna baboons and gorillas spend their entire lives within about 15 square miles,[9] and this is greater than the areas explored by most primates, which cover about four or five square miles.[10] But, in contrast, we cover significantly larger areas when we live as hunters and gatherers. For example, the Kung bushmen cover about 600 square miles in the course of their lives.[11]

Another study compared a variety of primates in terms of the number of years spent for socialization and play. Tree shrews, lemurs, tarsiers, squirrel monkeys, baboons, macaques, and chimpanzees were studied. The study demonstrated that the higher the level of a given primate species on the evolutionary scale, the longer the years spent for its socialization and play. Among primitive primates, object manipulation is absent and play is stylized. In contrast, advanced primates show less stylized behavior and a greater amount of object manipulation.[12] Yet another study indicates a clear association between evolutionary level, on the one hand, and increased object exploration and play, on the other.[13]

How can we explain the presence of stimulus seeking behavior among animals, and especially among advanced primate species? There are two possible explanations. First, it is possible to think that, as the primates evolved, this form of behavior developed more and more because to know the environment well is evolutionarily advantageous. Or, to put it differently, the primates evolved because of the development of the capacity to understand the environment better.

From the Darwinian perspective, we may say that individuals with a greater tendency for curiosity about the environment survived better and left more offspring with the tendency for this behavior pattern, resulting in this pattern's gradual dominance within the species. When a new species emerged, the same process continued, and its stimulus seeking behavior developed further. According to this explanation, stimulus seeking behavior is genetically given and innate.

Second, another possible explanation is that stimulus seeking behavior is primarily learned. If we take this position, we are assuming that, although animals may have an inborn tendency for curiosity, they basically acquire stimulus seeking behavior through learning. An animal learns stimulus seeking behavior because such behavior is rewarding.

Many forms of behavior, among both humans and other animals, are learned. Elephants, bears, horses, or chimpanzees in a circus are trained to perform very complicated behavior, because they know that they can obtain food as reward by doing so. Many human children do their homework in return for a promise of a reward, such as a new bicycle or a tennis racket. In these situations, a certain form of behavior is acquired through getting a reward for it.

In order to find out whether or not stimulus seeking behavior is innate or learned, we may look at several interesting experimental studies, which give us an unmistakable answer. One study deals with rhesus monkeys. In this study, the monkeys were exposed to a mechanical problem. The task, for instance, was to raise a hasp that is kept in place by both a hook and a pin. What a rhesus monkey can accomplish is to raise the hasp by removing the hook and pin. When this gadget was installed in the cage, the monkeys worked on it and solved it as many as seven or eight times over several days. The most important point in this study is that they received no reward whatsoever for solving this problem, yet they worked on it and solved it repeatedly.[14]

Another well-known study also deals with rhesus monkeys. In this experiment, a rhesus monkey was confined in a closed box. There were two doors with different cards placed on them. When one of the doors was pushed from the inside, the door opened, and the monkey could see toy trains moving in the adjoining room for 30 seconds. The monkeys in this experiment continued to open the right door in order to see the toy trains moving in the next room.[15] In this experiment, too, no food was given when the correct door was opened. The only thing the monkey could get was to look at the condition in the adjoining room. Interest in the environment, then, appears to be the only reason for doing so.

There are also studies dealing with chimpanzees. Many chimpanzees appear to enjoy "artistic" activities such as drawing and painting. They concentrate intensely on their work without any materialistic reward.[16] These studies collectively suggest that stimulus seeking behavior is genetically given to advanced animals, such as primates.

In order to avoid misunderstanding, I should add that the result of stimulus seeking behavior *per se* can be rewarding; by recognizing the consequence of one's own manipulatory behavior, or by being exposed to a new visual experience. The point I would like to make is that the animal does not learn to show stimulus seeking behavior purposefully in order to get a reward such as food. Stimulus seeking behavior is genetically given to the animal, and the consequence of such behavior is not the purpose of the behavior.

MANIPULATORY SKILL IN PRIMATES

The examples we have seen deal with rhesus monkeys and chimpanzees. Although we do not have a large number of detailed studies about the stimulus seeking behavior of many primate species, available studies clearly show an unmistakable tendency among primates—when compared with nonprimates—to actively seek to manipulate the environment, mainly by using their fingers in a sophisticated manner. For example, in her excellent study of the chimpanzees in the Gombe National Park in Tanzania, Jane Goodall repeatedly tells us about a variety of manipulatory skills the chimpanzees show.[17]

Perhaps the most famous example in her study is so-called "termite fishing." These chimpanzees love to eat termites, and they almost always use tools to get them. Tools are fashioned from grass, vines, bark, or twigs, and then carefully inserted into the passage of a termite mound. After a while, when the tool is withdrawn, termite soldiers may be found on the tool, and are eaten. This sounds simple. But both Jane Goodall and another chimpanzee researcher, Geza Teleki, found out how difficult it was to fish termites by trying it themselves. These two human beings could not even insert a tool into the passage, and when it was withdrawn, they did not get any termites at all. The secret was that a tool must be inserted into the passage gently by rotating it carefully, because the passage is not straight. Furthermore, when the tool is in the passage, it must be vibrated gently in order to attract termites. After that, the tool must be withdrawn smoothly and gracefully, so as not to separate the termites from their jaws. Chimpanzees are fully competent to do this manipulatory activity, which requires very minute coordination of the muscles of the fingers and arm.[18]

Goodall also tells us a variety of tool-using behaviors among the chimpanzees in Gombe. Leaves, grass, twigs, sticks, and rocks are used as a sponge, mop, brush, napkin, container, toy, missle, club, hammer, weapon, or fly whisk.[19] The chimpanzees' active manipulation of material objects is indeed impressive.

However, although they may be more skillful in the art of termite fishing than us, we are also good at manipulating our environment in different ways. What this makes us realize is that, through evolution, animals have become more and more active in their stimulus seeking behavior, and that this is based on the more purposeful and intentional manipulation of the environment. As psychologist Harry F. Harlow and others have suggested, it is quite reasonable to assume the existence of the "manipulatory drive" at least among the advanced primates such as chimpanzees and us.[20]

Furthermore, it is also possible to conceive of a physical space in which the manipulatory drive is expressed. Philosopher George Herbert Mead coined an interesting expression, "manipulatory area." By this he means a physical space consisting of objects that an organism sees and is able to handle. When the result of manipulating an object is experienced, the organism experiences its own distance to it. The manipulatory area is such a range of experiences in space.[21] If we assume the existence of the manipulatory drive and its expression in the manipulatory area among advanced primates, it may be logical to look for the biological and evolutionary basis for its origin and existence in the history of the primates.

THE EVOLUTION OF PRIMATES AND ARBOREAL THEORY

The theory generally accepted by both specialists and the general public regarding the evolutionary past of *Homo sapiens* is the so-called "arboreal theory," which was first systematically proposed in 1912.[22] According to this theory, the remote ancestors of the living primates were terrestrial creatures, which resembled today's tree shrew (a primitive primate species that superficially looks like a rat). This creature somehow began to have an arboreal way of life.

This theory was later popularized by British anatomist Le Gros Clark. He assumed that this ancestral animal climbed trees with clawed hands and feet, which were not prehensile. The eyes and the brain were small, but its olfactory apparatus was elaborate. Through time, however, various evolutionary changes took place in its anatomy because of its arboreal life. The fingers and toes acquired the ability to grasp tree branches. The five digits changed so that they could be splayed out and bent together in a converging movement, making the grasping of small objects possible. The thumb and big toe tended to increase their relative size and to become capable of being moved in opposition to the other digits for grasping purposes. This was further enhanced in most primates by the transformation of sharp claws into flattened nails. The development of the nails was accompanied by the emergence of highly sensitive pads at the tips of the digits, providing a very efficient grasping mechanism capable of delicate manipulations.

Le Gros Clark further argues that since arboreal life requires accurate judging of distance and direction, the development of good eyesight was necessary, and the visual organ became elaborate and suitable for arboreal life. The eyes were set close to each other to allow stereoscopic vision, which makes the judgment of distance easier. The demand for skill and cunning in arboreal life was one of the reasons why the brain began to expand in size and complexity in the evolutionary history of the primates.[23]

Others emphasized another plausible consequence of arboreal life. It has been noted that the chimpanzee, the gorilla, and our own species have wrists that are functionally very similar,[24] as well as laterally directed shoulder joints and remarkable mobility of the shoulder and elbow joint complexes.[25] Together with many other anatomical details common in advanced primates, these characteristics were explained as consequences of brachiation.

Brachiation refers to the movement of swinging from one branch to another by the arms. In carrying out brachiation, an animal hangs from a tree branch by one arm, swings, and reaches for another tree branch with the other arm and grasps it. It then hangs from this branch by freeing the other arm, which is then used for reaching yet another tree branch swinging. By repeating this sequence of movement, the animal can move among the trees without coming down to the ground. In addition to brachiation, the behavior of reaching in many directions while climbing and feeding in trees has also been pointed out as one of the reasons for the anatomical and structural similarities in the arms of the advanced primates.[26]

ARBOREAL THEORY AND THE ORIGIN OF THE MANIPULATORY DRIVE

Assuming for the moment that our ancestral primates actually did lead an arboreal life, how was this way of life instrumental in developing the manipulatory drive in the living advanced primates? A conceivable reasoning may be made as follows. The movement of brachiation involves the stretching of one arm to reach a tree branch and to hold it, so that the animal can swing in order to reach the next tree branch and hold it by stretching the other arm. This movement is likely to be accompanied by the evolution of the consciousness, in the animal, that it must reach for something actively. The brachiating animal is here intentionally extending the arm to reach for a tree branch, to hold it firmly, in order to use it as the basis for the next movement of brachiation. It is very likely that a strong drive to utilize the tree branch as an object of manipulation underlies the movement. Without such a drive, the animal is not likely to extend the arm and to reach for a tree branch.

The development of stereoscopic vision is also understandable in connection with the development of brachiation. Since brachiation requires the accurate judgment of the distance between tree branches, it is necessary to have a visual system that allows the animal to make such a judgment correctly. Stereoscopic vision is helpful for brachiation, because two eyes focused on the same object can achieve the purpose much more effectively and accurately than only one eye or two eyes not focused on the same object, as in the case of the horse, for example.

Intelligence is an advantage for an animal in any situation, and, in this context, the development of intelligence among the evolving primates is likely to help the animal make a judgment as to whether or not it should try to reach for a particular tree branch, and as to whether or not it should continue brachiating. To have adequate intelligence in this situation means that the animal can know the range and extent of what can be manipulated in the environment.

The evolution of the prehensile digits and opposable thumbs as well as the replacement of the claw with the nail and pad is a significant advantage for manipulatory behavior. When the five digits are very much alike and all move in the same direction, as in the case of the marmoset (a primitive primate species), the animal can merely hang on by its claws. But when the thumb is opposable and moves in the opposite direction compared with the other digits, the animal can hold an object between the thumb and the index finger and can grasp objects.

The development of the pad at the tip of the finger, accompanied by the concentration of sensory receptor neurons in the pad, has an additional advantage for manipulatory behavior, because the animal can now obtain more information about the object it is holding. With two hands of this kind, the animal can pick up a small object and hold it firmly and use it against another object either in the environment or in the other hand, held similarly between the thumb and index fingers. This totally new capacity in the evolution of the animal kingdom cannot be emphasized too much. The animal can now manipulate two objects in both hands at the same time, and, when one of the two objects changes or modifies the other, it functions as a tool. Without this possibility, efficient stone tools, for example, cannot be produced.

An important phenomenon in this connection regarding the evolution of primates is the general increase in body size over time. The primitive primates such as the tree shrew or the marmoset are as small as the rat or squirrel; the more advanced primates, however, such as the gorilla, the orangutan, the chimpanzee, and our own species, are much larger than these less evolved primates. Larger body size means on the whole longer arms, which in turn mean a greater space for manipulation. Also, larger body size itself means greater muscle power of the limbs, which also contributes to the establishment of a larger manipulatory area.

These changes are also likely to be accompanied by the development of a stronger drive to manipulate, because larger primates can indeed manipulate more powerfully and effectively than smaller primates, and, without a strong drive to manipulate, long and strong arms are meaningless. Only when the animal has a strong drive to manipulate does its long and powerful arms have an evolutionary meaning, and it is difficult to assume the existence of such arms without the corresponding drive to use them adequately and properly.

THE FORAGING-IN-THE-BUSH THEORY AND THE ORIGIN OF THE MANIPULATORY DRIVE

Although widely accepted, the arboreal theory is not free from criticism. For example, it has been pointed out that there are arboreal animals that can quite successfully survive on the trees without primate-like characteristics. The diurnal tree squirrel does not have close-set eyes. Its eyes face laterally, and the visual fields from the two eyes overlap rather poorly. All the digits bear claws. There are no opposable thumbs or toes. These features are, according to arboreal theorists, less adaptive to arboreal life and therefore are not found in advanced primates. Yet everyone knows that squirrels are highly skillful and successful in their arboreal movements.[27] This certainly is a decisive weakness of the arboreal theory. For this reason, there are those who reject the theory of the arboreal origin of the primates.

For example, M. Cartmill suggests that close-set eyes, grasping digits, and reduced claws may have been adaptations to a foraging way of life, with the animal seeking fruits and insects in bushes on the ground. The development of stereoscopic vision is adaptive for life on the ground because the animal can judge its distance from prey accurately without having to move the head. The claws were a hindrance in the bush, where the animal must grasp slender twigs, and therefore evolved into the grasping digits, because such digits allow the animal to hold securely onto thin supports when using both hands to catch the prey.[28]

According to this theory, which might be called the "foraging-in-the-bush" theory, then, the evolution of the opposable thumb, the nail, the pad at the tip of the digit, as well as stereoscopic vision, have been adaptative for catching insects, plucking fruits, and moving around in the bush by grasping slender supports.

Assuming this time that the primates evolved on the ground in the bush, rather than through the arboreal way of life, we may also recognize factors that possibly made the manipulatory drive adaptive in the primates. The hypothetical animal in this situation moves around in the bushes and shrubs, grasping slender twigs, most conceivably using all four limbs. It stretches its limb for a twig, holds it, shifts its center of gravity toward it while freeing one or more of the other limbs, and seeks a new twig by stretching another limb.

Exactly as in the case of brachiation, then, the evolution of the consciousness in the animal that it must reach for something actively is conceivable in this way of life, accompanied by the drive to manipulate. The behavior of plucking fruits is also likely to be instrumental in evolving the manipulatory drive. But, above all, catching live insects, which may escape if one makes the slightest mistake in moving the limbs, is likely to contribute to the evolution of the manipulatory drive. The animal is likely to evolve in muscular coordination of limbs and digits as well as in its active attempt or even determination to catch moving insects.

THE MANIPULATORY DRIVE IN *HOMO SAPIENS SAPIENS*

Whether the ancestral primates were living on trees or on the ground, then, there is good reason to assume that environmental and ecological factors could have been instrumental in their developing the drive to manipulate objects in their environment. It is often speculated that our ancestral tree-dwelling primates existed during the Oligocene period (about 38–26 million years ago), and one estimate by Vincent Sarich, based on molecular data, suggests that brachiation began about 12–10 million years ago.[29] These data suggest that there has been enough time for primates to evolve the drive to manipulate objects in their environment actively and intentionally.

We know that all living large apes, namely, the chimpanzee, the gorilla, the orangutan, and we ourselves show a very active attempt to manipulate objects. There are various estimates regarding the evolutionary divergence between the human line (hominidae) of evolution, on the one hand, and the gorilla and the chimpanzee (pongidae), on the other. The technique of chromosomal banding and gene mapping suggests that the divergence took place about 8–6 million years ago,[30] and immunological studies of albumin offer an estimation of about 8 to 4 million years ago.[31] The divergence of the orangutan is estimated to have taken place earlier than this.

Certainly, these estimates vary to some extent, but that should not bother us. We can say confidently that at least several million years have passed since the divergence. The presence of the strong manipulatory drive among the chimpanzee, the gorilla, and us can be a result of parallel evolution since then, but, in view of the surprising similarity in the ways this drive is expressed, as well as the anatomical similarities among us, it is far more logical and natural to assume that the strong manipulatory drive was already present in our common ancestry, and also even before the divergence of the orangutan. This further suggests that we have inherited a drive that can be traced back in our genetic background at least 10 million years.

Originally, the manipulatory drive was likely to be expressed in grasping a tree branch and in swinging in order to prepare for the next movement of "reach-and-grasp-and-swing," if we accept the arboreal theory. But this drive can be easily expressed in other forms, especially when the fingers are prehensile and the thumbs opposable. Highly developed intelligence helps and directs the animal to use the same drive to manipulate its environment in various ways other than reach-and-grasp-and-swing.

In all forms, however, when this drive is expressed, the principle is the same: Reach for an object, hold it, and utilize it for the animal's advantage. I think the manipulatory drive in all primates is always characterized by this principle, whether the animal in question is a gibbon, a chimpanzee, or a human being. Such a drive could have developed logically either from brachiating between trees or from foraging in the bush on the ground.

Here is a potentially fruitful assumption that we can make in understanding ourselves, our culture, and our behavior. By assuming that we have genetically acquired a strong drive to manipulate from our biological heritage as an advanced primate species, we can derive insight into a large variety of phenomena: for the individual, between individuals, within society, and between nations and cultures. By making this assumption, we may understand ourselves better at the personal, interpersonal, regional, national, international, and global levels in a new perspective. This is one of the two key assumptions in this book.

THE MANIPULATORY DRIVE AND AGGRESSION

In this book I assume that, through evolution, the primates have acquired very active stimulus seeking behavior, and that the drive to manipulate objects in the environment underlies such a pattern of behavior. This drive, namely the manipulatory drive, can be observed in basic forms of stimulus seeking behavior such as exploration, play, and creativity. In these situations, the animal's behavior is aimed at acquiring more information about the environment, interacting with the environment on the basis of the information thus acquired, and deriving novel effects through such interaction. Stimulus seeking behavior in essence is the behavior dealing with the animal's relations with its environment. For this reason, the manipulatory drive is both necessary and instrumental in carrying out stimulus seeking behavior.

However, the manipulatory drive may also become instrumental in carrying out other forms of behavior dealing with mating, hunting, collecting food, or caring for the young. These forms of behavior are by nature not stimulus seeking, but since the manipulatory drive makes the animal utilize its environment to its advantage by learning about, modifying, or changing it, such a drive also helps the animal to carry out these behaviors. To acquire a better knowledge about the environment may also result in the discovery of a prey or a mate, or the protection of the young from danger. In addition to these preliminary advantages, the manipulatory drive also helps the animal to carry out feeding by actually capturing a prey or copulation by making a mate submissive. The young may be protected by counterattacking and chasing off attacking animals. In these cases, the manipulatry drive is directly beneficial for feeding, mating, and reproductive behaviors.

Admittedly, it may be difficult to differentiate between the manipulatory drive and, for example, the sexual drive as the underlying force in the animal when we observe a mating behavior as a phenomenon. Nevertheless, there can be behaviors that can be interpreted as primarily based on the manipulatory drive. There are good descriptions of chimpanzee behavior that we can quote here to illustrate our point.

Temper tantrums are often observed among chimpanzees. Jane Goodall states that when a chimpanzee infant in Gombe seeks its mother's breast and is rejected twice by her, it may throw a tantrum. A nearby adult male may become irritated and become threatening by raising arms. By recognizing this, the mother may relent and comply with her child's demand.[32] R. M. Yerkes makes another observation: "I have seen a [chimpanzee] youngster, in the midst of a tantrum, glance furtively at its mother . . . as if to discover whether its action was attracting attention."[33] Goodall says: "The temper tantrum seems to be an uncontrolled, uninhibited, and highly emotional response to frustration. It appears as a last resort, when [a chimpanzee] failed to get his way. It is observed most frequently in youngsters who are going through the peak of weaning."[34]

As a chimpanzee child gets older, its play becomes increasingly rough and aggressive. It learns the power relationships among playmates, as well as among their mothers, and at the same time develops skills in fighting, bluffing, and forming alliances.[35] The struggle for dominance is one of the most common causes of aggression among adult male chimpanzees at Gombe.[36] Adolescent males try to dominate females first before they are mature enough to try to dominate other males.[37]

A mature male courts a female by a variety of signals such as gazing at her, shaking a branch, rocking, and stretching arms. When the female responds by approaching him, he gets up and moves away, looking back over his shoulder to make sure she follows. If she does not, he stops and repeats his gestures. If she still shows reluctance, his actions become increasingly violent. He may even attack her.[38]

Chimpanzees may attack and kill other animals without eating them. Two chimpanzees were observed to seize civet cats and beat and stamped on them until they were immobile. The victims were left there and later died of injuries.[39] Aggression of this nature may be directed against individuals of the same species in the same community. Killers may even return to the scene of their aggression, search for the victim, and inspect the dead body.[40] Goodall states: "It is as though the aggressors check up on the results of their attacks."[41] When chimpanzees are involved in a dispute over territory, in contrast to the relatively peaceful and ritualistic solution of the dispute found among many species, they not only repel intruders but sometimes injure and eliminate them as well.[42]

In these descriptions, we can recognize behaviors that can best be described as "manipulatory." The manipulatory drive dominates and overshadows all other behaviors in each case. Temper tantrums are seen in an infant chimpanzee when it cannot obtain milk, and often work as a means to get it. Adult males display aggressive behavior toward females in order to force them to copulate, and youngsters learn the display of aggression through play. Animals are killed not for the sake of eating them but for the sake of manipulating them. Death occurs because the chimpanzees destroy victims completely. They further make sure that victims have been killed—that is, completely manipulated—by returning to the place of violence. The same behavior is seen toward fellow chimpanzees both inside and outside the community.

In these examples, it is possible to recognize the result of the manipulatory drive; there are active and purposeful attempts to manipulate the environment in order to make it more desirable and acceptable to the individual chimpanzee. Aggression, then, is an extreme form of manipulatory behavior in which the manipulatory drive is expressed violently and powerfully. It is interesting to note that chimpanzees may fight just for the sake of fighting. In Goodall's words, they fight "for the fun of it."[43] Here, we can see a situation in which aggression is a form of play. Of course, mating or territorial behavior as such may genetically entail aggressive behavior, but, by having the manipulatory drive, the animal can carry out such behaviors more extremely and extensively, as these examples show us clearly.

STIMULUS SEEKING BEHAVIOR IN *HOMO SAPIENS SAPIENS*

Exactly as in the case of other advanced primates, *Homo sapiens sapiens* shows great interest in its environment in the form of exploration, play, and creativity. Psychologist R. S. Woodworth argues that a great deal of human behavior is directed toward producing effects upon the environment without the immediate servicing of any aroused organic need. To him, to deal with the environment *per se* is the most fundamental motivating element.[44] In our daily language exploration may be called "expedition," "excursion," "holiday trip," or "scientific investigation." But its nature is the same in all cases: an outlet for people trying to find out about their environment. We climb mountains, explore the antarctic, and send space ships to other planets and to outer space. We spend an enormous amount of resources merely to find out about our environment. In practice, most of the fields of knowledge pursued at universities and colleges are explorations in sophisticated forms.

When stimulus seeking behavior is directed toward other human beings, we meet new people at a party or invite people to our homes in order to "get to know each other." It is assumed, in terms of commonsense, that the sole meaning of parties and inviting guests for a dinner is to get to know people, and nothing else. Indeed, some people have stimulus seeking behavior excessively, so that they are considered "nosy" or "disturbing." What sociologist W. I. Thomas called "the wish for new experience"[45] in essence deals with the drive underlying the behavior of exploration.

In the case of play, we have behaviors that are very similar to the ones found among other primates, which we call "sports" or "games." In addition, much of our cultural activities belong to the category of play. Tools, machines, vehicles, and other material objects are utilized in order to maintain our daily activities. What sociologists call "institutions" are also "play" in terms of the primatological perspective, because institutions are the established and routine ways in which we deal with our physical and social environment.

Although creativity among the primates is generally limited, this form of stimulus seeking behavior has acquired great significance among us. Our "artistic," "scientific," and "scholarly" activities emphasize creativity to such an extent that often this is the sole meaning of these forms of stimulus seeking behavior. A work of fine art must be original and must create a new artistic experience among the audience in order to be considered a good work. A novel, poem, or drama also must be original if such a work is to be taken seriously. A parody is always considered second-

or third-rate work, even when it attracts the attention of the audience. Similarly, in the academic world of arts and sciences, only the original study is valued; the more original, the better. At the same time, stigma is placed on any work that is a copy or imitation of a work produced by someone else. To create a "novel effect" in the form of creativity has occupied the central position of stimulus seeking behavior among us, and this is the greatest difference between the other primates and us.

As in the case of chimpanzees, the manipulatory drive may become apparent from time to time in our behavior, often in a very bizarre manner. For example, it is not uncommon to see a competition in an "affluent" society in which the participant who has been able to eat or drink most or fastest becomes the winner. Wife-beating, child abuse, kidnapping, and hostage-taking for extortion of money or for a political or religious demand are overwhelmingly manipulatory in nature. Anthropologists Sherwood L. Washburn and C. S. Lancaster state that there are people "who use the lightest fishing tackle to prolong the fish's futile struggle, in order to maximize the personal sense of mastery and skill."[46]

But perhaps the most grotesque manipulatory acts are found in documents on the Nazi concentration camps. Witness the following example. One guard had a hobby of stopping prisoners just before they reached the latrine. He then forced a prisoner to stand at attention for questioning. He was then forced to squat in deep knee-bends "until the poor man could no longer control his sphincter and 'exploded.' " He was then beaten because of this and then allowed to reach the latrine.[47]

These examples are sufficient enough, I think, to show the nature of the manipulatory drive among us, which is observable in two forms: (1) it may be directed toward a definite goal, but (2) it may be expressed just for the sake of expressing it. In this second situation, an individual satisfies his or her manipulatory drive by merely expressing it. To gain or obtain something is not the main purpose of this second form of expressing the manipulatory drive. Satisfaction comes from releasing the manipulatory drive toward the environment. Here, successful manipulation *per se* is the source of satisfaction.

Both forms of manipulatory behavior are important to us, but I would like to emphasize the second form as an especially important element in understanding primate behavior in general and human behavior in particular. There are several approaches in the social and behavioral sciences in which human behavior is explained in terms of clearly goal-directed perspective. Commonly, such factors as sex, aggression, dominance, or power are often assumed to be the driving force underlying human behavior. These factors are no doubt important in understanding ourselves, but, at the same time, we must also recognize the aspect of our behavior that is not goal-directed.

This results from the second form of expressing the manipulatory drive. It is very important to remember that we do have this form of manipulatory behavior, but, for some reason, we tend to overlook or ignore this aspect of ourselves. But by paying our attention to it more carefully, we may be able to gain a better insight into our behavior and ourselves. For example, criminology, deviant behavior, and juvenile delinquency can be seen from a different perspective if we remember this aspect of ourselves.

In all forms of stimulus seeking behavior among us, the manipulatory drive is directed toward two kinds of objects: (1) material, physical objects, such as stones, trees, animals, and other human beings, and (2) symbols, such as ideas, concepts, words, and ideologies. Through the development of intelligence, we are capable of dealing with symbols as objects for manipulation, just like stones or sticks. This does not mean, of course, that our attempt to manipulate is always successful or that we can manipulate symbols as freely as we want. But our ways of dealing with symbols are often exactly as if we were faced with physical objects. For example, when we see a warning label on a box that says "Dynamite—Handle with care," we react to the words as if the words themselves were dynamite. In reality, the box may be empty.

Symbols are also like material objects in many situations. We can make people happy and euphoric by saying complimentary words, and, by making malicious and hostile remarks, we may make a person furious, feel depressed, or commit suicide. Quarrels, which are exchange of words in the form of symbols, may result in a duel between two men. In these situations, symbols are just like chemical compounds or weapons. To put it differently, people manipulate symbols and at the same time are manipulated by them, in the same sense that people manipulate material objects and are manipulated by them. Sociologist Erving Goffman tells us in detail how we manipulate each other symbolically in our social interactions.[48]

STIMULUS SEEKING BEHAVIOR AND CULTURE

When in the nineteenth century anthropologist E. B. Tylor made one of the first definitions of culture, he stated as follows: "Culture or civilization . . . is that complex whole which includes knowledge, belief, art, morals, law, custom, and any other capabilities and habits acquired by man as a member of society."[49] Here, it is very clear that he is referring to the culture of *Homo sapiens*, and anthropologists after him continued to think of culture as unique and limited to *Homo sapiens*.[50] But this belief has become more and more questionable in recent decades. Many studies of other primate species indicate that they, too, show behaviors that can be considered cultural.

For example, we know of Japanese monkeys that wash potatoes in sea water before eating them. In this way, presumably, they add a salty taste to the potatoes, and they appear to like it. They also have acquired a method of separating out wheat grains mixed with sand by dropping them in the water.[51] It is unpleasant even for Japanese monkeys to have sand in the meal, and by using this method, they successfully get rid of sand from the grains before they eat them. Another well-known example is termite-fishing among chimpanzees, which I mentioned earlier. Chimpanzees carefully and intentionally prepare tools in order to insert them into termite holes and fish termites out skillfully and patiently. In these examples, the animals learned the behavior; their behaviors were patterned, that is, stylized; and they shared the behavior exactly as in the case of human culture.

The phenomenon of culture is no doubt very complex, and it is not a simple task to define it. But if we focus our attention on the relationship between an organism and its environment, stimulus seeking behavior may be a key in understanding the basis for and origin of "culture" among us as well as other species. Along with the evolution of the primates, stimulus seeking behavior has become the more and more active manipulation of the environment, and, from this perspective, we may comprehend the nature of culture at least in part. Culture in this sense is an expression of stimulus seeking behavior in a more established form. Stimulus seeking behavior as well as its results become "cultural" when such behavior is learned, patterned, and shared, regardless of whether it is *Homo sapiens*, the chimpanzee, or the Japanese monkey that shows such behavior.

A question that arises at this point is this: How does an animal in this situation deal with its environment? What means does it use to mediate between its manipulatory drive and the environment? The answer is both a body of knowledge and the material objects to go with such knowledge, in the form of technology. Technology, then, is a means to achieve the objective of manipulation, and, accordingly, it is within the range of culture as conceived in the present context. By applying technology to the environment, no matter how primitive the technology may be, an organism can try to manipulate it, and when successful, it can establish a range of control, which lies within the bounds of culture for this particular species. Furthermore, the new knowledge obtained in this range of control can contribute to the improvement of technology to make manipulation more efficient and powerful.

Along with evolution, the manipulation of the environment has become more effective as well. The technique of manipulation among us has become sophisticated to such an extent that we tend to overlook manipulatory activities among other species. But when we look at the nature of culture in terms of the evolutionary perspective in the form of stimulus seeking behavior, it is more logical to accept that culture is not limited to us.

THE ARISTOTELIAN SCHEME OF TECHNOLOGY

Even though the phenomenon of culture is found in other species, however, human culture is unique in the sense that symbols and technology are extremely important. Symbols are objects to be manipulated, and technology is the means for manipulation. In reality, technology may involve more diversified aspects than we usually think. I would like to elaborate on this point by borrowing, modifying, and freely reinterpreting Aristotle's idea of technology.[52]

Aristotle's discussion of "cause" in his *Physics* is often interpreted from the perspective of technology. For example, Robert W. Daly thinks that there are six elements in Aristotle's scheme. These are: (1) the tool-user, (2) the tool, (3) the technique for using the tool correctly, (4) the material upon which the tool is used, (5) the intended goal or outcome, and (6) the result.[53]

For example, when I try to crack a nut in order to eat it, I am the tool-user, and the nutcracker is the tool. The technique for using the nutcracker is the knowledge I acquired through observation, and the material upon which the nutcracker is used is a nut. The intended goal or outcome is to crack the nutshell (but not the nut) and eat the nut inside the shell. The result may be successful, and I may be able to eat the nut as I wanted, but I may fail and I may smash both the nutshell and the nut into small pieces and I may have to sort out small pieces of nut before I can eat it.

If we think of a generalized situation in which a person as the subject is trying to do something by using a tool, the tool-user (element 1) is the self, and the technique for using the tool correctly (element 3) is the knowledge and skill the self has in using this tool. But the other four elements in the scheme, namely, the tool, the material, the goal, and the result, can be variable. Each of these four elements can be symbolic or materialistic. A materialistic object can be either animate or inanimate. An animate object can be either human or nonhuman. Therefore, the tool, the material, the goal, and the result may be drawn from any one of the following three kinds: (1) symbols, (2) inanimate or animate objects, or

(3) human beings. Of course, to classify the four elements in the Aristotelian scheme of technology into these three categories is purely heuristic.
The only reason for doing so is to understand the meaning of technology
in human culture better.

When the tool is symbolic, we have such examples as a political slogan,
a religion, an ideology, or simply a verbal attack or malicious remark. By
expressing these ideas as tools, one can work on the material. When the
tool is an inanimate object, we have such examples as machines, weapons,
or sticks. Penicillin is a tool that is animate. Mediators and messengers are
human tools.

The material upon which the tool is applied can be the opposing
ideology, another religion, or any different views. These are all symbols
that the tool-user might want to work on, influence, or change. Practically
any inanimate object can be the material to work on, and examples of
animate objects as materials are bacteria and mice in laboratories upon
which experiments are conducted. Although biological experiments on
humans may not be possible in many situations due to legal and ethical
reasons, human beings are nevertheless bombarded by means of symbols.
Atheists are told to believe in God, and radicals and conservatives try to
influence each other.

The goal or outcome of the use of the tool is, for example, equality,
freedom, utopia, or genocide, depending on the situation in which it is
symbolic. The goal or outcome can be a material object, as in building of
a house or landing on the moon. The creation of huge mice by means of
genetic engineering is an example involving animate objects. When the
goal involves human beings, there are such examples as brainwashing or
extermination of an ethnic group.

The result of tool-using may or may not agree with the goal of
tool-using, and, in certain cases, the result may be the same as the
intended goal. But, in other cases, the result may be significantly
different from the intended goal. The ideology of communism as a
symbolic system produced "counterrevolutionaries" within communist
countries. The problem of pollution and environmental destruction is to
a large extent due to the unexpected consequence of tool-using: that is,
the gap between the goal and result of tool-using. The fear of uncontrollable organisms made possible by genetic engineering touches upon
the possibility of the unintended result. The emergence of rebels and
alienated people in industrial society is again due to the gap between the
goal and the result. These are all examples of so-called "serendipity,"
involving consequences that are symbolic, inanimate or animate objects,
or human beings.

NOTES

1. M. J. Ellis, *Why People Play* (Englewood Cliffs, N.J.: Prentice-Hall, 1973), pp. 83–110.

2. R. Fagen, *Animal Play Behavior* (New York: Oxford University Press, 1981); and G. M. Burghardt, "On the Origins of Play," in *Play in Animals and Humans*, ed. P. K. Smith (Oxford: Basil Blackwell, 1984), p. 8.

3. C. Hutt, "Exploration and Play in Children," in *Play, Exploration and Territory in Mammals*, ed. P. A. Jewell and C. Loizos (London: Symposium of Zoological Society, 1966), pp. 61–81; Konrad Lorenz, *The Foundation of Ethology* (New York; Springer Verlag, 1981); W. I. Welker, "Ontogeny of Play and Exploratory Behaviors: A Definition of Problems and a Search for New Conceptual Solutions," in *Ontogeny of Vertebrate Behavior*, ed. H. Moltz (New York: Academic Press, 1971), pp. 171–228.

4. F. E. Poirier and E. O. Smith, "Socializing Functions of Primate Play Behavior," *American Zoologist*, 14 (1974), pp. 275–87.

5. J. D. Baldwin and J. I. Baldwin, *Beyond Sociobiology* (New York; Elsevier, 1981), pp. 184–85.

6. M. E. Berger, "Population Structure of Olive Baboons (Papio anubis) in the Laikipia District of Kenya," *East African Wildlife Journal*, 10 (1972), pp. 159–64.

7. O. Aldis, *Play Fighting* (New York: Academic Press, 1975); I. Eibl-Eibesfeldt, *Ethology: The Biology of Behavior* (New York: Holt, Rinehart and Winston, 1975); W. I. Welker, "An Analysis of Exploratory and Play Behavior in Animals," in *Functions of Varied Experience*, ed. D. W. Fiske and S. R. Maddi (Homewood, Ill.: Dorsey Press, 1961), pp. 278–325.

8. E. W. Menzel, Jr., "Chimpanzee Utilization of Space and Responsiveness to Objects: Age Differences and Comparison with Macaques," *Proceedings of the Second International Congress of Primatology*, 1 (1969), pp. 72–80.

9. George B. Schaller, *The Mountain Gorilla: Ecology and Behavior* (Chicago: University of Chicago Press, 1963); Irven DeVore and K.R.L. Hall, "Baboon Ecology," in *Primate Behavior*, ed. I. DeVore (New York: Holt, Rinehart and Winston, 1965).

10. Sherwood L. Washburn and C. S. Lancaster, "The Evolution of Hunting," in *Man the Hunter*, ed. Richard B. Lee and Irven DeVore (Chicago: Aldine, 1968), p. 297.

11. Richard B. Lee, "Subsistence Ecology of Kung Bushmen" (unpublished Ph.D. dissertation, University of California, Berkeley, Department of Anthropology, 1965).

12. A. Mazur and L. S. Robertson, *Biology and Social Behavior* (New York: Free Press, 1972).

13. D. K. Candland, J. A. French, and C. N. Johnson, "Object Play: Test of Categorized Model by the Genesis of Object-Play in Macaca fuscata," in *Social Play in Primates*, ed. E. O. Smith (New York: Academic Press, 1978), pp. 259–96.

14. H. F. Harlow, "Mice, Monkeys, Men, and Motives," *Psychological Review*, 60 (1953), pp. 23–32; H. F. Harlow, M. K. Harlow, and D. R. Meyer, "Learning Motivated by a Manipulatory Drive," *Journal of Experimental Psychology*, 50 (1950), pp. 228–34.

15. R. A. Butler, "Incentive Conditions Which Influence Visual Exploration," *Journal of Experimental Psychology*, 48 (1954), pp. 19–23.

16. K. Beach, R. S. Fouts, and D. H. Fouts, "Representational Art in Chimpanzees," *Friends of Washoe*, 3 (1984), pp. 2– 4; 4 (1984), pp. 1– 4; R. A. Gardner and B. T. Gardner, "Comparative Psychology and Language Acquisition," in *Psychology: The State of Art*, ed. K. Salzinger and F. E. Denmark (New York: New York Academy of Sciences, 1978); D. Morris, *The Biology of Art* (London: Methuen, 1962).

17. Jane Goodall, *The Chimpanzees of Gombe: Patterns of Behavior* (Cambridge, Mass.: Belknap Press, 1986).

18. Goodall, *The Chimpanzees of Gombe*, pp. 536–39; Geza Teleki, "Chimpanzee Subsistence Technology: Materials and Skills," *Journal of Human Evolution*, 3 (1974), pp. 575–94.

19. Goodall, *The Chimpanzees of Gombe*, p. 537.

20. Eibl-Eibesfeldt, *Ethology*, p. 276; Harlow, "Mice, Monkeys, Men, and Motives"; Harlow, Harlow, and Meyer, "Learning Motivated by a Manipulatory Drive."

21. George H. Mead, *The Philosophy of the Act* (Chicago: University of Chicago Press, 1938), p. 14.

22. G. E. Smith, *The Origin of Man*, Smithsonian Institution Annual Report, (Washington, D.C.: Smithsonian Institution, 1912), pp. 553–72.

23. W. E. Le Gros Clark, *History of the Primates* (Chicago: University of Chicago Press, 1959), pp. 46–53.

24. O. J. Lewis, "The Hominoid Wrist Joint," *American Journal of Physical Anthropology*, 30 (1969), pp. 251–68; O. J. Lewis, "Evolution of the Hominoid Wrist," in *The Functional and Evolutionary Biology of Primates*, ed. R. H. Tuttle (Chicago: Aldine, 1972), pp. 207–22.

25. R. H. Tuttle, "Knuckle-Walking and the Problem of Human Origins," *Science*, 166 (1969), pp. 953–61.

26. R. H. Tuttle, "Darwin's Apes, Dental Apes, and the Descent of Man: Normal Science in Evolutionary Anthropology," *Current Anthropology*, 15 (1974), pp. 390–97; S. L. Washburn, "Primate Studies," in *Nonhuman Primates and Medical Research*, ed. G. H. Bourne (New York: Academic Press, 1973), pp. 472–73.

27. M. Cartmill, "Rethinking Primate Origins," in *Primate Evolution and Human Origins*, ed. Russell L. Ciochon and John G. Fleagle (Menlo Park, Calif.: Cumming Publishing Co., 1985), p. 15.

28. Cartmill, "Rethinking Primate Origins," p. 18.

29. V. Sarich, "A Molecular Approach to the Question of Human Origins," in *Primate Evolution and Human Origins*, ed. Russell L. Ciochon and John G. Fleagle (Menlo Park, Calif.: Cumming Publishing Co., 1985), p. 321.

30. Brunetto Chiarelli, "Chromosomes and the Origin of Man," in *Hominid Evolution: Past, Present and Future*, ed. V. Tobias (New York: Alan R. Liss, 1985), pp. 397–400.

31. J. E. Cronin, "Apes, Humans and Molecular Clocks: A Reappraisal," in *New Interpretations of Ape and Human Ancestry*, ed. R. L. Ciochon and R. S. Corruccini (New York: Plenum Press, 1983), p. 134; Vincent Sarich and Allan Wilson, "An Immunological Timescale for Hominid Evolution," *Science*, 158 (1967), p. 1202.

32. Goodall, *The Champanzees of Gombe*, p. 115.

33. R. M. Yerkes, *Chimpanzees: A Laboratory Colony* (New Haven, Conn.: Yale University Press, 1943), p. 30.

34. Goodall, *The Chimpanzees of Gombe*, p. 576.

35. Ibid., p. 355.

36. Ibid., p. 325.

37. Ibid.

38. Ibid., p. 457.

39. Geza Teleki, "Notes on Chimpanzee Interactions with Small Carnivores in Gombe National Park, Tanzania," *Primates*, 14 (1973), pp. 407–12.

40. Goodall, *The Chimpanzees of Gombe*, p. 530.

41. Ibid.

42. Ibid., p. 528.

43. Ibid., p. 330.

44. R. S. Woodworth, *Dynamics of Behavior* (New York: Holt, 1958).

45. W. I. Thomas and Florian Znaniecki, *The Polish Peasant in Europe and America*, 5 vols. (Chicago: Universtiy of Chicago Press, 1918–20); W. I. Thomas, *The Unadjusted Girl* (Boston: Little, Brown, 1923).

46. Washburn and Lancaster, "The Evolution of Hunting," p. 299.

47. Alexander Donat, *The Holocaust Kingdom* (New York: Holt, Rinehart and Winston, 1965), p. 178.

48. Erving Goffman, *The Presentation of Self in Everyday Life* (Garden City, N.Y.: Doubleday, 1959).

49. E. B. Tylor, *Primitive Culture* (London: John Murray, 1871), p. 1.

50. A. L. Kroeber and C. Kluckhohn, *Culture: A Critical Review of Concepts and Definitions* (Cambridge, Mass.: Peabody Museum of American Archaeology and Ethnology, 1952).

51. M. Kawai, "Newly Acquired Pre-Cultural Behavior of the Natural Troop of Japanese Monkeys on Koshima Islet," *Primates*, 6 (1965), pp. 1–30.

52. W. Charlton, ed., *Aristole's Physics, Books I and II* (Oxford: Clarendon Press, 1985).

53. Robert W. Daly, "The Specters of Technicism," in *Alienation: Concept, Term, and Meanings*, ed. Frank Johnson (New York: Seminar Press, 1973), p. 341.

2

TO BE A HUMAN CHILD

The drive to manipulate the environment, which I assume to exist among us due to our evolutionary background, can also be inferred by examining the growth of human children. The ontogenetical consideration of *Homo sapiens sapiens* can offer us another basis for assuming the existence of the manipulatory drive among us.

THE GROWTH OF THE MANIPULATORY DRIVE

The Swiss psychologist Jean Piaget has made a significant contribution to our understanding of the growth and development of human children, including the development of the manipulatory drive.[1] According to him, in the first few months of an infant's life, objects play no instrumental role in his or her activities. Behavior is reflexive in the beginning, and it develops into the repetition of actions for their own sake. At this stage, play consists of the simple pleasure of functioning, and it is not directed toward objects. A child sucks, for example, in order to suck.

At three or four months, a change takes place. The infant becomes interested in manipulating its environment. For example, the child tries to prolong interesting sights by repeating kicking in order to make a mobile move. The child knows that by using its own body in the form of kicking, it can continue to make the mobile move. This is clearly the stimulus seeking behavior called play. The child begins to apply a variety of manipulatory movements toward its environment, such as shaking, banging, hitting, or rubbing objects nearby. However, the child's interest lies in the movement of the objects themselves; to make them move *per se* is interesting, and the child is not yet interested in using objects in order to achieve a certain objective, such as using an object as an instrument to move another object.

In the next stage, when the child is about one year old, it begins to differentiate between means and ends. For example, a child can use a stick in order to obtain an object.

As the child acquires the ability to deal with symbols, most conspicuously through the use of language, its relationship with the environment becomes more and more elaborate. The child can imagine a certain object as if it were something else, and can pretend to have it instead. The child can imagine a situation of having an object that it does not have in reality and behave accordingly. For example, a child may pick up a rock and consider it to be a puppy and play with and talk to it as if it were a real puppy. Another feature of symbolization is that simple physical experience may begin to mean a more abstract concept. For example, the physical comfort of a soft pillow may mean the quality of a living person, such as the mother.[2]

In the beginning, symbolization is rather limited, in the sense that the child does not or cannot use any object in order to symbolize. Most typically, a child uses dolls for symbolizing human beings instead of using toy cars or teddy bears. But, as the child gets older, symbols can be quite arbitrary, and the physical quality of the object that stands for a symbol becomes insignificant. It even becomes unnecessary that the object physically exists in front of the child.[3] The child becomes capable of thinking, behaving, and responding in a totally imaginary situation without any object to symbolize. Here, written or spoken words as symbols are enough to place the child in a position to manipulate symbols as if they were substantial, physical objects.

By examining the growth and development of human children, then, it is possible to conceive of how each human being acquires the skill to deal with his or her environment, which includes both physical and symbolic objects to manipulate. Although, because of the influence of culture as well as individual differences, it may be difficult to estimate the intensity of the manipulatory drive among human children in objective, quantitative terms, casual observation of tantrums among children in various cultures suggests that the manipulatory drive may be indeed very strong for them. Often children cry, scream, move their arms and feet violently, or refuse to listen to what others say when they cannot get what they want or when their attempt to reach for a material object is prevented by an adult. As I have mentioned in Chapter 1, temper tantrums in chimpanzee infants are violent, but human children show the same behavior at least equally violently.

THE EMERGENCE OF THE SELF AS OBJECT

Along with the development of the manipulatory drive, the human child gradually acquires the awareness of the self. The psychologist and philosopher William James discusses the two aspects of the self: the "I" and the "me." The "I" is that aspect of the self that is the subject, and the "me" is a different aspect of the self that is the object. The "I" as the subject is, according to James's terminology, the "pure ego," and the "me" as the object is called the "empirical self."

The empirical self in its widest sense is the sum total of all the person can call his or hers. It consists of three forms: the material self, the social self, and the spiritual self. The material self includes the person's own body to begin with, and this is surrounded by circles of material things, such as clothes, the family, home, and so on. What one has collected or created is also a part of the material self. Any material object that a person thinks of and feels as "his" or "hers" belongs to the material self.

The social self refers to the recognition a person receives from others. Insofar as a person cares about the opinion of another person, a particular kind of the social self exists. This means that there are as many versions of the social self as there are distinct individuals or category of individuals whose opinions are considered important by the "I," or the "pure ego." The spiritual self refers to a person's psychic faculties and collection of his or her states of consciousness.[4]

In James's distinction between the pure ego and the empirical self, then, the material self can become the object of the pure ego, or the "I." As a result, it may become the object for manipulation. A person manipulates or tries to manipulate material objects that are considered his or hers, such as a tool, a car, a business firm, or a family member such as a child or spouse. In addition, one manipulates one's own body as an object just like any other object recognized by the pure ego. The material self in this situation is an object in the form of a human being that has the unique property of having a special tie with the pure ego. The spiritual self is an exclusively symbolic object for manipulation, and the social self lies between the other two forms of the empirical self in this regard, having both materialistic and symbolic qualities as an object for manipulation.

EVOLUTION AND THE PARENT-CHILD CONFLICT

Conflict between parents and children can be understood in two ways. First, conflict is due to the incompatibility between the manipulatory drive of two individuals who happen to be the parent and the child. As individuals belonging to *Homo sapiens sapiens*, both parent and child are likely to have genetically acquired a strong drive to manipulate, and conflict between them is inherent in the genetic background through the evolution of the primates. Furthermore, due to the possibility of using symbols as objects for manipulation, they experience opposition to the attempt to manipulate both in the physical and in the symbolic worlds.

For example, both father and son may want to drive the only car of the family, and each may want to drive the car to a totally different place, such as a golf course versus a rock concert. At the more symbolic level, the father may want to listen to Bach in the living room, but the son may play such loud rock music that the father feels disturbed at home. In the first example, two individuals are trying to manipulate the same object for two different purposes, and, in the second example, one wants to manipulate symbols that the other does not want manipulated. Here, the possible source of conflict lies in the difference in taste and interest based on the difference in age and generation, but the conflict in this form is not limited to the relations between parent and child. Any two individuals can potentially experience conflicts of this nature.

Second, another source of conflict between parents and children can be traced back to the evolution of *Homo sapiens sapiens* in a somewhat different way. Through evolution, we turned out to be a species in which the parents are extremely concerned with the survival of their offspring. This point can be recognized clearly if we look at the evolution and change in the method of reproduction.

In the evolution of the phylum Chordata, one must first look at how fish reproduce themselves. Although there are species such as the well-known sticklebacks, which attend to and protect their eggs, most species produce an extremely large number of eggs and leave them alone. As a result, many of the eggs are eaten by other species. Those eggs that survive and turn into fish again are faced with the danger of being eaten. The ecological balance for fish species is based on this extremely high rate of attrition: The small portion of the eggs laid that survives is adequate to keep the species from becoming extinct.

In the case of the amphibians, the next form of life in the evolution of the phylum Chordata, the situation is not much better. Although they can leave the water temporarily and utilize oxygen in the air in order to survive, their reproduction is still connected to the water. The eggs are laid and left in or near the water, and the young begin life in the water. For this reason, they are faced with the same danger as fish.

In the next phase of evolution, which involves the reptiles, life has shifted from water to land, and reproduction is adjusted to the new way of life. The egg has acquired a hard shell, and this is an improvement compared with the soft, Jell-o-like egg of the fish and amphibians. In many species, the eggs are covered with sand before they hatch, and this also is an improvement for better survival and reproduction.

However, the emergence of the mammals marks a significant and decisive improvement in the effectiveness of reproduction. For the first time in the history of the phylum Chordata, the young are kept completely within the body of its mother until they are born. This is the safest possible way to maintain the growth of the young, for the young now cannot be eaten by an individual of another species, unless the mother itself falls a victim to that individual. The danger of being consumed as an egg in the early stage of life, which characterizes fish, amphibians, and reptiles, has been eliminated among the mammals. The child of a mammal species comes out of the protective body of its mother only when it is significantly close to the mature form of its species. (Certainly, we are an exception, and this point will be discussed shortly.)

Furthermore, after birth, the child is taken care of by its mother actively, and, again, this is a significant improvement over the earlier orders in terms of the better survival of the young and, accordingly, the better survival of the species. Unlike the lower orders of life, the number of offspring per mother is much less, and, among the most advanced primates, the mother gives birth to one offspring at a time. This form of reproduction allows the mother to pay her full attention to it. This would be totally unthinkable and unimaginable if there were as many offspring as in the case of a fish species!

Through evolution, then, fewer offspring are born, but more of them survive to reproduce. The method of reproduction has become more efficient by reducing attrition between birth and maturity. This improvement has been achieved by having the young protected in the mother's body among the mammals, and also by taking care of the young more actively through the evolution of maternal behavior.

BIPEDALISM, PREMATURE BIRTH, AND MATERNAL PROTECTION

This generalization holds in the evolution of the fish, amphibians, reptiles, and mammals. But in the case of *Homo sapiens*, there is a peculiar phenomenon associated with its evolution. That is the problem of bipedalism, or the mode of locomotion in which only the two hind legs are used, as we usually do. This way of locomotion, instead of using all four limbs as is the case for practically all nonhuman primates most of the time, resulted in the change in the pelvis. By walking upright, the human female has developed a narrower birth canal. To complicate the matter further, the fetus of *Homo sapiens* has acquired a larger brain in proportion to the rest of the body. This resulted in difficulties for the mother as the fetus passes through the narrow birth canal. The evolutionary solution is that the human baby is born earlier, quite prematurely compared with other mammal species, so that the size of the head is still small enough to go through the birth canal.[5]

This means that, compared with the newborn young of almost any other primate species, the human baby is far more helpless and dependent on its mother for survival. The behavior of maternal care is certainly not completely genetic, but it must be learned both among humans and nonhuman primates to a considerable extent.[6] But the human female has genetically acquired the propensity to take care of her own young upon being exposed to such behavior through socialization and learning. The human mother learns to be a human mother by activating the propensity for maternal behavior that she has acquired genetically.

This genetic propensity for maternal behavior may or may not be activated, depending on the mother's experience. But if we look at the fact that in practically all cultures and subcultures known to us in the past and at present, the majority of mothers show the maternal behavior of nursing, protecting, and raising their children, one sees that, in most cases, this propensity *is* activated. Of course, the method of taking care of children may not be satisfactory in terms of a certain perspective, such as the contemporary Western liberal values, but at least children survive, and that is the point. Such a helpless creature as a newborn human baby cannot possibly survive without a significant investment of parental care. Therefore, we may infer that human females have inherited a strong genetic propensity to take care of the young, which is to be activated through socialization and learning. It is difficult to activate a form of behavior that is not genetically prepared among human females of almost all known cultures and subcultures.

Conceivably, there are two reasons to infer such a genetic propensity. First, as an advanced primate species, the human female has acquired the drive to take care of the young effectively by paying attention to it more than do less evolved species. It is interesting to learn that, according to Goodall, one of the causes of mortality among chimpanzee infants is injuries resulting from falling from the mother.[7] No doubt this is much less common among us as a cause of infant mortality than among chimpanzees, and human mothers are more careful in handling their babies.

Second, because the human young, compared with the young of other mammal species, are relatively premature at birth as a result of bipedalism and the increased size of the brain of the fetus, it was necessary for *Homo sapiens* to develop more intense care-taking behavior to protect its helpless offspring. Without the evolution of such behavior, the helpless young are much less likely to survive, which in turn means the danger of extinction for *Homo sapiens* as a species.

Thus, it is reasonable to assume the existence of a strong drive in the female of *Homo sapiens* to protect the young and raise them carefully. This is by no means a mechanical and automatic matter. There are mothers who abuse, abandon, desert, and even kill their own children. Yet we have been one of the most successful species in terms of reproducing ourselves and spreading to a wide variety of geographical environments. This fact alone may be a good reason to infer that the strong protective drive among human mothers is in most cases activated successfully in order to deal with the helpless young.

In a sense, the drive to protect and to raise the young is not too dissimilar from the drive to manipulate. In both situations, one visualizes a certain physical space and assumes that a certain animate or inanimate object within this range is to be grasped, manipulated, and controlled. In the case of brachiation, a tree branch is recognized as an object within reach, and, on the basis of this assumption, the animal reaches for it and grasps it. When the animal swings and reaches for the next tree branch, it has manipulated the first tree branch for its own purposes, and the manipulation has been successful.

In the case of maternal behavior, the mother recognizes her child as an object within her own space of manipulation and control. When the child tries to move too far away from her, she prevents it from doing so; it is physically grabbed as an object of manipulation and kept close to her. It is under control. The picture changes considerably as the child gets older and older, but in this earliest phase of life, the mother's protective behavior is similar to her manipulatory behavior, such as brachiation. From this point of view, it is possible that the protective drive might have developed relatively quickly in response to bipedalism and premature birth on the basis of or by utilizing the genetic material for the drive to manipulate.

A SOURCE OF PARENT-CHILD CONFLICT

Here is, then, a source of conflict between parent and child. On the one hand, the human baby is born with a strong drive to manipulate, which begins to unfold more and more as it grows. But, on the other hand, the human mother has a drive to protect and control the child. In the beginning of the child's life, there is not much of a problem because the child is totally helpless and entirely dependent on its mother. The mother's protective drive predominates on the whole of the mother-child relationship. As the child gets older, however, its manipulatory drive comes into more and more conflict with the mother's protective drive. The child may not be allowed to reach for, touch, or manipulate objects. In the beginning the objects may vary from a bottle of vitamin pills to scissors; later they may be such objects as the motorcycle and the narcotic.

When the child reaches the stage when symbols begin to have significant meaning, conflict may arise because the teenage son joins a new religious cult or a radical organization. In these situations, the child wants to reach for a material or symbolic object. The manipulation of such an object may be manual, such as opening the bottle of pills, or it may be symbolic, such as reading, understanding, and evaluating a doctrine. The human mother in such a situation thinks that she must prevent her child from being exposed to a danger that may result in a body damage or injury or in a symbolic handicap such as social isolation or stigma. Both consequences are certainly undesirable in terms of the child's successful survival in human society.

In the case of the human male, understanding of his relationship with the child is the reason for exercising his manipulatory drive over it. The knowledge that his sexual act, which is in part a form of manipulatory behavior, resulted in the birth of the child is based on language as a systematic collection of symbols, and, once having this knowledge, he can look at his child as an object for manipulation as well. That is, a successful sexual act with a female is play as a form of stimulus seeking behavior. Sexual experimentation with the partner may be in some cases adequate enough to be considered as creativity as another form of stimulus seeking behavior. Certainly, when the child is born, this is a novel effect caused in part by the father's sexual behavior, falling within the category of creativity. The interaction between the father and child is also creative because new experience is derived from it.

As an object that is the result of his own act, the child, according to the father's assumption, can be manipulated and controlled as he wants. In the beginning of the child's life, the father can do so without much resistance from the child. But, exactly as in the relationship between mother and child, the father's relation with the child becomes more difficult as the child gets older, due to its own drive to manipulate. Two different forms of the drive to manipulate crash and collide between father and child, involving material and symbolic objects.

Compared with other primate species, the parent-child relationship among us has become further complicated by the fact that the mother's protective behavior lingers on much longer than is necessary for the survival of the child. This point can be illustrated by looking at the mother-child relationship among other primates.

For example, a study of the common Indian langur shows a sharply different picture. Phyllis Jay, a primatologist, studied this primate species in India in the late 1950s and kept an interesting record of its behavior. Until it is about a year old, the langur enjoys a wonderful life, as far as its relationship with the mother is concerned. The mother is permissive and protective. But when the child is about a year old, it is time to be weaned, and this happens quite abruptly and unbelievably to the child. The mother suddenly begins to refuse to let the child suckle. Not only that, she clearly shows rejecting behavior. When the child runs after her, she runs away. If the child continues to follow her, she covers her teats. She may even slap the child if it still demands to suckle.

The infant throws tantrums, reminding us of a human child. Also like a human child, it screams and slaps the mother. No matter what the infant does, however, the mother simply ignores it. She may also appear to be irritated. This relationship may continue several months, and the infant may be as big as, or even bigger than, the mother. In the meantime, the mother becomes pregnant again, and the child's happy and pleasant relationship with the mother is gone forever.[8]

Here, what can be considered a form of parent-child conflict is totally different from the kind found among us, except for certain cultures in which weaning is also enforced abruptly and severely. In the case of the common langur, conflict occurs because the mother terminates her maternal behavior even though the child wants to have it continued. Among us, on the contrary, conflict arises because the mother continues to show protective behavior when the child does not want it and begins to reject it. Not only that, in most cases and in most cultures, the mother's concern about her child continues probably for life, and, to many children, this may be experienced frankly as annoying.

I am not saying that human mothers should be like langur mothers. I am not sure which is better. The points I would like to make here are, first, that the langur mother and child experience conflict differently from us because the mother terminates her maternal behavior when the child still wants to have it, and, second, that the human mother and child experience conflict mainly because the mother wants to continue her maternal behavior when the child does not want it any longer.

These points are important in evaluating sociobiologist Robert L. Trivers's view on parent-child conflict. According to him, maternal rejection at weaning among many species can be seen as an attempt by the mother to cut her investment in her genes. By weaning the infant and by giving birth to the next child, she can leave more offspring and, accordingly, more of her genes. Therefore, he reasons, to terminate the care of the young early and to prepare for the next child is advantageous from the evolutionary perspective.[9]

This is a conceivable possibility, but it is difficult to apply this theory to *Homo sapiens sapiens*. Admittedly, there are cultures in which the infant, almost like the langur infant, is rejected by the mother as soon as she becomes pregnant again, a well-known example being the Alorese on the island of Alor in Indonesia.[10] But, in worldwide context, an extreme culture like this is very rare.[11] Furthermore, it is difficult to apply this explanation to cultures in which weaning is late, slow, and gradual. Trivers argues that there has been an evolutionary change in the infant, who uses the psychological technique of prolonging nursing, by smiling toward the mother, for example.[12] But this sounds like a heads-I-win-tails-you-lose argument. It looks like the claimed evolutionary mechanism of investing genes effectively by weaning early and producing more offspring suddenly failed when *Homo sapiens sapiens* emerged.

THE MANIPULATORY DRIVE AND THE PARENT-CHILD RELATIONSHIP

Psychoanalyst Erik H. Erikson has formulated a well-known scheme of personality development, and reinterpretation of his idea from our perspective may help us in further understanding the nature of parent-child conflict.[13] According to Erikson, personality develops throughout a person's life, and there are eight stages of personality development. Each of these eight stages is characterized by two alternatives, and a person goes through one of these at each stage, depending on his or her relationship with other human beings. At each stage, there is a new dimension in a person's growth. It is interesting to note that the first four stages deal with conflict between parent and child, involving the drives of protection and manipulation that we have just discussed.

The first stage, according to Erikson's scheme, has the alternatives of what he calls "trust versus mistrust," and this applies to the child's first year of life. At this stage, whether the child learns to trust or mistrust its environment depends on the quality of care it receives. If the child receives good care during the first year of life, it begins to understand that the world is a safe place and people are helpful and dependable. But if the care the child receives is inadequate or rejecting, it learns to mistrust the world. As a result, fear and suspicion develop in the child's mind.

At this stage, during the first year of life, the child is totally dependent on the mother or a mother-substitute, and the issue is whether or not the mother expresses her maternal drive to protect and raise the child satisfactorily toward the child. The child's manipulatory drive is not yet strong enough to become a serious problem of incompatibility with maternal care. It should be added that, according to Erikson, the issue of trust versus mistrust may not be resolved once and for all during the first year of life. A child who has learned to mistrust people at this stage may begin to trust people by meeting reliable people at a later stage. Similarly, a trusting child may learn to mistrust people at a later stage by going through a traumatic experience, such as the divorce of its parents.

The next stage is called "autonomy versus doubt and shame," and the child is at this stage when it is in the second and third years of life. Here, the child's autonomy begins to develop, and, as a result, conflict between the mother's protective drive versus the child's manipulatory drive, in the sense we defined them, emerges. According to Erikson, this is the stage of developing motor and mental abilities, and the child actively walks and climbs. If the child is allowed to move as it wants, it develops a sense of autonomy. The child realizes that it can control its muscles, itself, and its environment. On the other hand, if the caretaker does too much for the child, for example, by reaching for something the child wants to reach and by giving it to the child, it learns to doubt its ability to control. If the caretaker becomes impatient in a situation like this, the child acquires the feeling of shame. To criticize bedwetting, soiling, spilling, or breaking objects may also make the child feel ashamed.

In terms of the analytical scheme of the two opposing drives assumed in this book, then, autonomy is produced when the child's manipulatory drive is allowed to be expressed. When the mother prevents this by exercising her drive of maternal care, the result is doubt on the part of the child. When the mother does not allow for mistakes in connection with the exercise of the manipulatory drive such as spilling or breaking, the child becomes ashamed. Both are negative responses to the child's drive to manipulate.

The next stage, "initiative versus guilt," applies when the child is four and five years old. The key point at this stage is how the parents respond to the child's self-initiated activities. If the child is free to initiate motor play such as running, bike-riding, or skating, the child's sense of initiative is reinforced and supported. The child realizes that its attempts to explore and play with the environment and to derive novel effects out of it—in brief, its stimulus seeking behavior—are approved and encouraged. At this stage, the child's curiosity also becomes symbolic through the development of language ability, and if the parents answer the child's questions seriously without avoiding or rejecting them, the child learns that its attempts to initiate action are good and bring about positive results. In contrast, if the child is made to feel that its motor activity is bad or its questions are a nuisance, a sense of guilt is implanted in the child's mind.

The fourth stage is "industry versus inferiority" and applies to when the child is between six and eleven years old. The child at this stage is now capable of deductive reasoning. The child knows that there are rules to follow in order to play a game. If the child is encouraged to make, do, or build something and is praised and rewarded for what it has created, the child acquires the sense of industry. The child learns the pleasure of creating something. In contrast, if such activities are seen by the caretaker as mischief or making a mess, a sense of inferiority develops in the child's mind.

These first four stages are predominantly under the influence of the caretakers, usually the mother, and the second, third, and fourth stages are about conflict between the two forms of drive as expressed by the parent and child. In Erikson's scheme, personality development after the fourth stage turns to deal with situations outside the family, and the two alternatives in each of the remaining four stages are symbolic; a person is faced with two alternatives of predominantly symbolic situations. Since we are dealing with the parent-child conflict here, I shall not discuss the last four stages here.

It is certainly true that, as a child gets older, the focus of interest shifts to activities outside the family, and the stimuli for personality growth and alternative situations lie outside the family. Nevertheless, a person does not terminate his or her contact with family members, and the parents continue to show concern over their children even when they have become adults. Conflict due to two opposing drives continues to linger on in many situations of a person's adult life.

IDENTIFICATION AS AN ADJUSTMENT MECHANISM

How can a conflict of this nature be dealt with? It may be impossible to avoid conflict completely insofar as people are different from each other in terms of their genetic background, experiences in life, and position in a group or in society. The relationship between parent and child is no exception; these two people are not the same and have different wishes and drives. We can easily understand that if two individuals, a parent and child, should attempt to realize their drives fully as much as they want, the result would be disastrous. In view of the strong manipulatory and protective drives we have, we may wonder if we have also a built-in biological mechanism to deal with such a situation, which can at least reduce the extent and intensity of conflict between the two opposing drives.

It is conceivable that the psychological and psychoanalytic concept of identification refers to such a mechanism. Identification means the process and result of placing oneself in someone else's position through imagination and understanding. By doing this, a person can obtain insight into how another person thinks, feels, and looks at a variety of situations. In this way, a person may understand why and how the other person is trying to exercise his or her manipulatory drive in ways that appear objectionable and offensive.

The evolution of such a mechanism in the mind is most adaptive for a species in which a strong manipulatory drive is present. Indeed, identification and the manipulatory drive are likely to have evolved more or less concurrently. A strong manipulatory drive would be detrimental to the species if there were no mechanism for identification, because individuals would end up exterminating each other. At the same time, there would be no need for identification if individuals did not try to exercise the manipulatory drive. In such a situation, a person is not interfered with by another person, and there is no need to understand how others think and feel.

It is very interesting to discover that the chimpanzee, another species with a strong manipulatory drive, is also capable of identification. The chimpanzee researcher D. Premack conducted a series of experiments with a chimpanzee named Sarah. She was shown videotaped scenes of a human being faced with a variety of problems to be solved.

For example, the human actor was trying to escape from a locked room or was shivering when an electric heater was unplugged. After the video, Sarah was then shown a pair of photographs of each scene, and one of the pair showed a solution to the problem, such as a photograph of a key, or of a plugged-in heater. Surprisingly, Sarah consistently chose the correct photographs. In a later study, she was exposed to a more difficult task: After the same series of videofilms, she had to choose among three alternatives of photos, such as intact, bent, or broken keys; or a plugged-in heater, an unplugged heater, or a plugged-in heater with a cut electric cord. Sarah again chose photos correctly and consistently.[14]

The most logical explanation for these results is that Sarah was able to look at the problem from the standpoint of the person in the video. She was able to point out the correct thing to do, such as using an intact key to open the door, or plugging in the heater, in view of a situation that was not a problem for her but rather for the filmed person. This is clearly an example of identification. There are also studies showing identification between chimpanzees, that is, a chimpanzee identifies with another chimpanzee.[15]

In view of these experimental findings, the fact that a chimpanzee mother almost always gives in to her child's temper tantrum for milk[16] strongly suggests that she identifies with her child and understands its strong need for feeding. As a matter of fact, the chimpanzee mother is quite different from the Indian langur mother. As you will recall, the langur mother mechanically ignores her child's demand for milk when the time for weaning comes. This difference between the chimpanzee and the langur reflects the difference in the level of evolution as to the ability to identify between these two species.

Coming back to the discussion of the human mother and child, we may state the following: By means of identification with the mother, the child may be willing to reduce the exercise of the manipulatory drive toward the mother, and the mother in turn may try to understand the child's need to manipulate. Of course, this is a matter of degree, but if there were no identification at all, the relationship between mother and child would be a constant battle based on physical power.

Especially for the child, the ability to identify with the mother, father, and other caretakers, and with adults as a whole, helps it to minimize its miserable realization of powerlessness vis-à-vis adults. Identification allows the child to pseudo-experience the exercise of its drive to manipulate by seeing an adult manipulating objects. The child vicariously manipulates objects and may be able to obtain the satisfaction of having manipulated something itself. This becomes possible when the child has acquired enough ability to symbolize through the development of intelligence, and, in terms of the psychoanalytical perspective, this is possible after the resolution of the Oedipus conflict, in which the child learns to identify with the parent of the same sex.

IDENTIFICATION AS A MECHANISM WITHIN SOCIETY

The child further continues to identify vicariously with a wide variety of individuals who can impress it by manipulating something splendidly. For example, a child identifies with professional athletes, who can manipulate a ball or their own bodies successfully to such an extent that the child as an observer can derive the satisfaction of manipulation. A child aspires to drive a truck or locomotive or to fly an airplane in the future. A child may also be fascinated by a dancer, circus artist, musician, or magician—who can show the skill of manipulation marvelously—and thus satisfy its manipulatory drive vicariously.

In the early stage of the child's identification, it is easier to identify with someone whose exercise of the manipulatory drive is visible and directed toward material objects rather than invisible and symbolic ones. At such a stage, the child does not identify with a lawyer, mathematician, composer, or philosopher, who manipulates symbols almost exclusively, and whose manipulatory behavior therefore is more difficult to comprehend.

When a child becomes older and is able to understand symbolic manipulation in society such as the expression of power and power struggle, identification may be directed toward individuals who can manipulate more at the symbolic level. Individuals may identify with political, religious, or financial leaders in society, accept their views, and even worship them. As in the case of a child identifying with its own parent, identification in situations like these also helps to reduce potential conflicts between people. Imagine the chaos possible if everyone in society were to try to exercise their manipulatory drives in the way he or she wanted! By having people identify with and follow a leader, a society can avoid constant struggle and conflict over the exercise of power as a symbolic expression of the manipulatory drive. In this sense, in terms of the evolutionary perspective, the development of ability to identify with someone who manipulates at both materialistic and symbolic levels is logical and understandable, in view of the very strong manipulatory drive that we have acquired as a primate species.

When a person vicariously experiences his or her manipulatory drive by identifying with someone who actually exercises it in one way or another, satisfaction appears to come mostly from seeing and hearing the ongoing manipulation. In the world of popular music, a very interesting phenomenon is known. Often successful musicians who have established themselves commercially through records are imitated by local performing musicians, and their success in turn depends on the faithfulness of their imitation.[17] It may be difficult to understand why anyone would want to listen to the performance of imitators when the original is available on records. But one conceivable answer is that by nature, we want to experience someone else's manipulation as it actually takes place in reality. Live performance as an ongoing act of manipulation entails the thrill of seeing or hearing the manipulation that is actually taking place at this moment, not in the past. This experience of seeing and hearing a live performance makes the vicarious experience of the listener more exciting and intense.

Since musicians are not machines operated by a prepared program, each performance is especially created only on that occasion, and this is one source of the attraction of a live performance. The conductor Antal Dorati makes the following remark: "One of the greatest attractions of a musical performance . . . is the element of improvisation. . . . No two performances are alike. We not only know that, we also expect it; we look forward to the inevitable novelty at the next hearing of the same piece by the same performer or performers."[18]

Psychologist D. O. Hebb tried to explain this phenomenon of "difference-in-sameness" neurologically.[19] As Dorati points out, this is the difference between live music and recording, because recorded performance is always exactly the same.[20] From a record, you cannot expect to derive a new experience. A person's manipulatory drive is not something frozen in the past but a biological force alive in his or her body, and, in order to satisfy this vicariously, a performance done in the past on record has a significantly reduced effect as to identification.

Exactly the same remark is applicable to other forms of entertainment such as sports, the circus, play, or magic, as well as identification with a political, religious, or financial leader. The persuading force of a leader is most powerful in the "live" situation, and an old film of a charismatic leader such as Hitler is hardly persuading.

IDENTIFICATION WITH THE AGGRESSOR

Since a child learns to experience vicariously the exercise of the manipulatory drive in order to satisfy its own drive, identification is more likely to be directed toward someone who can effectively manipulate objects. Depending on the nature of the objects to be manipulated, there are two kinds of persons one can identify with. First, there are individuals who have the competence, ability, or skill to manipulate symbols or material objects effectively. To a child's eye, professional athletes, truck drivers, musicians, and magicians are examples of this. Second, there are also individuals who can manipulate other human beings effectively even against their will. They can exercise power over others, and, in extreme situations, human beings are manipulated exactly like inanimate objects. A child may identify with such a person with power by merely realizing the possibility of being manipulated by him or her.

This phenomenon has attracted the attention of Sigmund Freud, who conceived of this as the core of father-son conflict, known in psychoanalysis as the Oedipus conflict. According to Freud, the son loves the mother and sees the father as a rival. But by realizing the father's power, the son identifies with the father. In this way, he can vicariously experience his incesteous love for the mother through the father's position in the family and at the same time reduce his fear of the father's power by thinking and feeling like the father by means of identification. This is the solution of the Oedipus conflict.[21]

Anna Freud, Sigmund Freud's youngest child, studied the phenomenon of identification with a powerful person; she called it "identification with the aggressor."[22] According to her, identification with the aggressor can be observed in a variety of situations. She talks about the case of an elementary school boy who had a habit of making faces when he was scolded by his teacher. The teacher thought that either the boy was consciously making fun of him or else had an involuntary tic. A psychoanalytical examination revealed that the boy identified himself with the teacher when he was scolded and imitated the teacher's expression involuntarily.

Anna Freud gives another clinical example, that of a girl afraid of ghosts who suddenly came up with a splendid idea for how to overcome her fear. When she had to cross the hall at home in the dark, she would make all kinds of strange gestures. She explained her behavior as follows: "There is no need to be afraid in the hall. You just have to pretend that you are the ghost who might meet you." By making strange gestures, she herself became a ghost, and, in this way, she had no reason to be afraid of ghosts. By impersonating the aggressor in the form of a ghost, she transformed herself into the aggressor, eliminating her fear of being threatened.

According to Anna Freud, identification with the aggressor is a normal phenomenon for a child, and she considered this to be "a preliminary phase of superego development."[23] A well-known experimental study in social psychology corroborates this point.[24] Interestingly, even adults may identify with the aggressor when they are forced to think and behave like children, as in the case of the prisoners in the Nazi concentration camps and the Japanese in occupied Japan after World War II.[25]

Some aspects of so-called initiation ceremonies commonly found in the anthropological literature may be explained from the same perspective. Lionel Tiger and Robin Fox, two anthropologists, suggest that initiation ceremonies in various cultures are conducted in order to control the rebelliousness of the younger generation.[26] Such ceremonies for the young often include various arrangements such as circumcision, hazing, symbolic death and rebirth, acquisition of a new status, or learning a societal secret. Tiger and Fox think that the adult members of the society handle the rebellious youth by taming them, making them submissive, and forcing them to identify with the adult members of society. Their reasoning can be supported by the fact that younger members of widely different societies all over the world are almost universally required to go through formal or informal initiations before they can acquire a full status as an adult member of society. In terms of the assumption made in this book, we may state that initiation ceremonies result from two opposing drives to manipulate held by the older and younger generations. This is, so to speak, an institutionalized way of dealing with parent-child conflict collectively by means of identification with the aggressor.

NOTES

1. Jean Piaget, *Play, Dreams, and Imitation in Childhood* (London: Routledge and Kegan Paul, 1951); Jean Piaget, *The Origin of Intelligence in Children* (New York: International Universities Press, 1952).

2. Anna Freud, *Normality and Pathology in Childhood* (New York: International Universities Press, 1965); D. W. Winnicott, *Playing and Reality* (London: Tavistock, 1971), p. 4.

3. Piaget, *Play, Dreams, and Imitation in Childhood*; N. Overton and J. Jackson, "The Representation of Imagined Objects in Action Sequences: A Developmental Study," *Child Development*, 44 (1973), pp. 309–14.

4. William James, *The Principles of Psychology*, vol. 1 (New York: Dover Publications, 1950), chap. 10.

5. S. J. Gould, *Ontogeny and Phylogeny* (Cambridge, Mass.: Belknap Press, 1977), p. 370; Helen E. Fisher, *The Sex Contract: The Evolution of Human Behavior* (New York: Quill, 1983), pp. 82, 220.

6. H. Harlow, "The Development of Affectional Patterns in Infant Monkeys," in *Determinant of Infant Behavior*, ed. B. Foss (London: Methuen, 1961), chap. 19; H. Harlow, "The Heterosexual Affectional System in Monkeys," *American Psychologist*, 17 (1962), pp. 1–9.

7. Jane Goodall, "Mother-Offspring Relationships in Chimpanzees," in *Primate Ethology*, ed. D. Morris (Chicago: Aldine, 1967), pp. 287–346.

8. Phyllis Jay, "The Common Langur of North India," in *Primate Behavior: Field Studies of Monkeys and Apes*, ed. Irven DeVore (New York: Holt, Rinehart and Winston, 1965), pp. 197–249.

9. Robert L. Trivers, "Parent-Offspring Conflict," *American Zoologist*, 14 (1974), pp. 249–64.

10. Cora DuBois, *The People of Alor: A Social-Psychological Study of an East Indian Island* (Minneapolis, Minn.: University of Minnesota Press, 1944).

11. Ronald P. Rohner and Evelyn C. Rohner, "Parental Acceptance-Rejection and Parental Control: Cross-Cultural Codes," *Ethnology*, 20 (1981), pp. 245–60.

12. Trivers, "Parent-Offspring Conflict," p. 257.

13. Erik H. Erikson, *Childhood and Society*, 2d ed. (New York: W. W. Norton, 1963), chap. 7.

14. D. Premack and G. Woodruff, "Does the Chimpanzee Have a Theory of Mind?" *Behavioral and Brain Sciences*, 1 (1978), pp. 515–26.

15. Jane Goodall, *The Chimpanzees of Gombe: Patterns of Behavior* (Cambridge, Mass.: Belknap Press, 1986), pp. 36–37.

16. Ibid., p. 576.

17. Kurt Blaukopf, "New Patterns of Musical Behavior of the Young Generation in Industrial Societies," in *New Patterns of Musical Behavior of the Young Generation in Industrial Societies*, ed. Irmgard Bontinck (Vienna: Universal Edition A.G., 1974), pp. 18–19.

18. Antal Dorati, *Notes of Seven Decades* (London: Hodder and Stoughton, 1979), p. 299.

19. D. O. Hebb, *The Organization of Behavior* (New York: Wiley, 1949).

20. Dorati, *Notes of Seven Decades*, p. 299.

21. Sigmund Freud, *On Sexuality* (London: Penguin Books, 1977), pp. 315–22.

22. Anna Freud, *The Ego and the Mechanisms of Defense*, rev. ed. (New York: International Universities Press, 1982), chap. 9, especially pp. 110–11.

23. Ibid., p. 120.

24. Ralph White and Ronald Lippitt, "Leader Behavior and Member Reaction in Three 'Social Climates,' " in *Group Dynamics*, 2d ed., ed. Dorwin Cartwright and Alvin Zander (London: Tavistock, 1960), p. 545.

25. Bruno Bettelheim, "Individual and Mass Behavior in Extreme Situations," *Journal of Abnormal and Social Psychology*, 38 (1943), pp. 417–52; Michio Kitahara, *Children of the Sun: The Japanese and the Outside World* (New York: St. Martin's Press, 1989), pp. 93–96.

26. Lionel Tiger and Robin Fox, *The Imperial Animal* (London: Secker and Warburg, 1971), pp. 158–59.

3

DEVIANT BEHAVIOR

The problem of deviant and criminal behavior in society has been explained in a variety of ways by sociologists, psychologists, and psychoanalysts. Understandably, their explanations are plausible and thus deserve serious attention in explaining deviance and crime. However, relatively little attention has been paid to our primatological background in understanding deviant and criminal behaviors. I would like to look at these problems within such a perspective in this chapter.

DEVIANCE AS STIMULUS SEEKING BEHAVIOR

Simply put, deviant behavior can be seen as behavior in which stimulus seeking behavior has become excessive or unusual. When such behavior is against the law, it becomes crime. In order to explain this point, it is possible to recognize four characteristics often found in deviant behavior, which are also the common characteristics of stimulus seeking behavior.

First, a person experiences a strong "urge" to do it. For example, a 14-year-old boy says that he "wanted to 'split tires' and did . . . meanness, I guess—get an urge to do it—start with one and keep on doing it. . . . Well, it didn't matter, any car would do. . . . . Teenagers . . . feel [the] urge to do something ornery."[1] Similarly, a man describes how he became a professional gambler. When he was 25, he went to a fashionable nightclub with his friends. He did not drink, and he knew very little about such places. There was a gambling room, and out of boredom and curiosity he drifted over to a crap table and watched the game. While he watched, he got a sudden impulse to play, and, before he could check himself, he was in the game.[2] These stories clearly tell us that a person is almost driven into such behaviors.

Second, often a person experiences a "kick" or "thrill" by attempting a deviant act. One obtains satisfaction by manipulating objects in the environment exactly as in play, but this behavior happens to be considered undesirable or illegal. A boy in a gang describes his own experience as follows. When he was eight, he followed other boys who were seven or eight years older. They went to a butcher shop to break in one night. They could open the transom, but it was too little for the older boys to get through. He was thrilled when he was told to crawl through the transom. He describes this experience in his own words as follows: "That was the kick of my whole life. . . . I was too thrilled to say no."[3]

Discussing the activity of an adolescent peer group in a Boston slum, sociologist Herbert Gans states: " 'Action' generates a state of quasi-hypnotic excitement which enables the individual to feel that he is in control, both of his own drives and of the environment."[4] It is clear that this description is applicable to any kind of play as a form of stimulus seeking behavior. Similarly, in a study of adolescent gangs in Chicago in the 1920s, another sociologist, Frederic Thrasher, recognizes thrill, excitement, and new experience as important elements in or reasons for the gangs' behavior.[5]

In the case of motorcycle gangs, it is theoretically conceivable that the source of the "kick" lies in the rapid movement of the body. However, although this element may not be excluded, it is more likely that a youngster derives satisfaction from three sources: (1) manipulating both his own body and the motorcycle successfully despite the risk of failing to do so (exactly as in the case of a circus artist), (2) the realization that one is violating a societal norm, and (3) attracting the attention and negative responses of adults. If speed were the only source of excitement, this could be equally well fulfilled by riding a roller coaster, but youngsters do not do this, presumably because (1) the roller coaster is safe to ride, (2) one cannot manipulate it when one rides it, and (3) it is socially accepted behavior.

Stealing, at least among boys, can be seen as an attempt to experience "kicks." According to sociologist George Grosser, boys steal all sorts of things, and frequently the stolen goods are of no use to them. They steal just for "fun." Girls steal things they can use, such as clothing, jewelry, and cosmetics.[6] The fact that these American boys steal indiscriminately whether or not they can use the stolen goods is in accordance with the greater tendency to desire power, money, and prestige in America among men than among women, regardless of whether this is biological or cultural.

The possibility for failure, no matter how slight it might be, seems to be present in other forms of deviant behavior as a source of thrill, excitement, or kick. For example, voyeurism, or "peeping," is a source of sexual excitement when it involves the risk of being discovered. A male voyeur is excited by his anticipation of how the woman may react if she finds out that he is watching.[7] A voyeur cannot obtain satisfaction when the woman knows from the beginning that she is being watched; one possible explanation for this is that by realizing that the behavior of "peeping" has been successful despite the danger of being discovered, a voyeur feels himself to be in complete control of his stimulus seeking behavior. The perfect stimulus seeking behavior makes him realize that he has manipulated the situation well. The same pattern of deriving a thrill and excitement has been reported in case studies of pickpocketing.[8]

In the case of gambling, the situation is reversed. One well knows that one usually does not win in a gamble. Yet one is told that one may, if one is lucky, win a fortune. In this situation, the possibility for winning appears to be a powerful encouragement in making one try to manipulate a very difficult situation successfully. This is the same as achieving an especially hard task, and, in this way, a person acquires a sense of successful manipulation, and this is a form of play.

A gambler derives intense excitement in the anticipation of winning a game. A professional gambler describes this as follows: "I stood around a few moments and then started to play. I was very excited; my hands were so wet you'd think I was washing them, and my heart was going a mile a minute."[9] Edmund Bergler states that "pleasurable-painful tension" or "thrill" is experienced by the gambler between the time of betting and the outcome of the game, and without the ingredient of "thrill" no game is worth betting to a gambler. Bergler thinks this is one of the characteristics of the gambler.[10]

Third, deviant behavior is often enjoyable. It is committed because it is fun. A boy describes the activities of his gang as follows:

We did all kinds of dirty tricks for fun. We'd see a sign, "Please keep the street clean," but we'd tear it down and say, "We don't feel like keeping it clean." One day we put a can of glue in the engine of a man's car. We would always tear things down. That would make us laugh and feel good, to have so many jokes.[11]

Sometimes rules for the game develop. A boy describes his behavior:

I would go into a store to steal a cap, by trying on one and when the clerk was not watching walk out of the store, leaving the old cap. With the new cap on my head I would go into another store, do the same thing as in the other store, getting a new hat and leave the one I had taken from the other place. I might do this all day and have one hat at night. It was fun I wanted, not the hat.[12]

Deviant behavior can become creative. The same boy describes a game his gang played:

When we were shoplifting we always made a game of it. For example, we might gamble on who could steal the most caps in a day or who could steal in the presence of a detective and then get away.[13]

In these examples, these delinquent boys enjoy their experience through deriving a novel effect in their stimulus seeking behavior, and this is a form of creativity.

Fourth, excessive stimulus seeking behavior is seen among those who are addicted to a wide variety of material objects and gadgets, including such modern innovations as the computer and the video game. Here, instead of relying on the chance factor, one can, through practice and effort, improve one's skills for manipulation, and this becomes the source of satisfaction. Whether or not one can manipulate the object satisfactorily is entirely up to one's skill, which can be improved by effort.

These four characteristics are commonly found in stimulus seeking behavior. If such behavior is directed toward objects in a socially and culturally accepted way, one is praised for one's hard work, effort, diligence, and achievement. Achievements in the arts, science, technology, business, or religion are in most cases accepted, and when one achieves the goal of manipulating material or symbolic objects successfully one is rewarded for that through honor, money, or prestige. Addiction to one's activity is universally found in any significant achievement. In contrast, when the effort is directed toward objects that are socially not accepted, or when the stimulus seeking behavior itself is not accepted, the behavior is considered deviant. When such behavior is against the law, one becomes a criminal.

Among these four characteristics, the playlike nature of deviant behavior is quite often found in juvenile delinquency.[14] Indeed, if we limit our consideration only to juvenile delinquency, its parallel to play as a stimulus seeking behavior becomes even more evident, in two ways. First, both play and juvenile delinquency are typical of children and juveniles. Play is generally considered to be a characteristic of immature rather than adult animals.[15] It is mainly these pre-adult animals, including human children and juveniles, that engage in play. It is known that delinquent behavior among human juveniles tends to disappear when they turn to a conventional way of life as they become mature adults.[16]

Second, both play and juvenile delinquency are more common among males than among females. Males play harder, begin play earlier, and cease play at a later age among primates.[17] Crime and delinquency rates for human males are consistently and significantly higher than those for females for all age groups—except when such deviant forms of behavior as prostitution, infanticide, and abortion are criminal. This generalization applies to all nations, to all communities within a nation, and to all periods of history for which organized statistics are available.[18] If we look at these parallel features in play and deviance in general and play and juvenile delinquency in particular, we have good reason to infer the element of play in deviant behavior.

Creativity is also an important element in criminal behavior in general. According to several studies, criminals look at the normal physical environment from a new perspective, so that burglary, for example, can be committed successfully.[19] A professional criminal states as follows: "Lots of the things I've sat down and figured out myself."[20]

DEVIANCE, CRIME, AND THE ARISTOTELIAN SCHEME

This way of looking at deviance and crime can be further elaborated by using the Aristotelian scheme of technology, which we discussed earlier, in Chapter 1. As you will recall, in the approach suggested by Robert W. Daly, an individual's behavior toward his or her environment in this scheme is analyzed in terms of (1) the tool-user, (2) the tool, (3) the technique for using the tool correctly, (4) the material upon which the tool is used, (5) the intended goal or outcome, and (6) the result. Since it is assumed that both deviant and criminal behaviors are special forms of stimulus seeking behavior to which the Aristotelian scheme of technology is applicable, we may examine these behaviors in terms of this scheme here. Deviant and criminal behaviors can be understood from a new perspective on the basis of this scheme.

The Tool User. The category to which the tool-user belongs may or may not make his or her behavior deviant or criminal. For example, in Sweden in the past, a woman who smoked in the street was deviant, although this behavior was acceptable for men. In contrast, even today, a man who carries a handbag is deviant, but this behavior is considered normal for a woman. In the American deep South in the past, an Afro-American was to be arrested for violating the law by merely entering a restaurant to be served or by waiting for a train or bus in a waiting room not intended for his or her racial group. A passenger holding a second-class ticket is not permitted to travel in the first-class coach of a train: in this case, however, in most situations this passenger is considered neither deviant nor criminal.

The Tool. Regardless of the user, the fact of having a certain tool itself may become a source of deviant or criminal behavior. To have a snake in the living room is deviant, and to possess a machine gun can become a crime.

The Technique for Using the Tool Correctly. When one uses a tool in a way other than the one for which it is intended, the act of doing so may become deviant. To live in a car instead of driving it for the purpose of transportation is deviant, and so is the use of a swimming pool for breeding fish. Reckless driving can be either deviant or criminal, depending on how serious the driving infringements are.

The Material. In some cases, to conceive of a certain object as a material to reach may mean deviance. Toward the end of the fifteenth century, to try to reach "India" by sailing west from Europe was certainly deviant in thought. This was because "India" was not considered to be within the range of the manipulatory area as seen westward from Europe, and, in this sense, "India" did not exist in that direction. Even when the object itself is known to exist as a reality, the manner of carrying out stimulus seeking behavior toward it may make a person deviant. When the material to be worked on is legally owned by someone else, the result of manipulating it may become criminal, such as joy-riding or stealing a car.

The Intended Goal or Outcome. By definition, this element refers to the idea held by the tool-user, but even in this situation deviance and crime may be seen. To organize a new religious sect, revolution, or political activity such as a right-wing or left-wing extremism is deviant from the norm of society, and, in some countries, to plan such an activity itself can be a crime.

The Result. Most deviant or criminal behaviors, however, are judged as such by looking at the result of action. This means that when the intended goal or outcome does not materialize, no deviance or crime is recognized. But, at the same time, the result that was not at all intended may be judged to be deviant or criminal behavior. For example, a doctor may be sued for malpractice for an event he or she may not at all have intended to bring about.

It is clear, then, that the possibility for deviant or criminal behavior can exist in all of the six elements in the Aristotelian scheme of technology, which we assume to be applicable to any situation of stimulus seeking behavior. Among these six elements, however, it is especially important to note the fourth element, that is, the material to which the tool is applied. As we discussed earlier, the material in this sense can be quite varied, and for the sake of convenience we have recognized three major categories: (1) symbols, (2) inanimate or animate objects, and (3) human beings. I would like to elaborate on deviance and crime by focusing our attention on these three forms of objects in the Aristotelian scheme.

WHITE-COLLAR CRIME

Deviant or criminal behaviors involving symbols are found in many forms in contemporary, industrial societies. So-called "white-collar crimes" are exclusively symbolic in nature. Criminologist Edwin Sutherland published the result of his study dealing with crimes committed by 70 large corporations and their subsidiaries over a period of several decades. The crimes were misrepresentation in advertising; infringements of patents, trademarks, and copyrights; financial fraud; and violation of trust. These crimes were committed by businessmen in connection with their work. These 70 corporations had an average of four convictions each.[21]

Business in the contemporary industrial world depends almost entirely on the manipulation of symbols—for example, figures in the balance sheet and reports to shareholders, taxation offices, and auditor's statements. It may also rely on the use of language in the statements describing the condition of the corporation to shareholders, governmental authorities, and the general public in the form of advertisements. On the manufacturing side of the corporation, symbols may include mathematics, statistics, chemistry, physics, medicine, engineering, biology, psychology, and so on, in all of which a variety of symbols are manipulated.

These symbols are highly specialized, and nonspecialists cannot easily comprehend them. Sutherland states that in most cases of white-collar crime the general public does not change its image of these large corporations despite their criminal behavior, and one of the conceivable reasons may be due to the layman's difficulty in understanding in what way and how a white-collar crime is committed.

VANDALISM

The second category of deviance and criminal behavior deals with inanimate or animate objects, and, as a good example involving inanimate objects, vandalism may be mentioned. According to criminologist Stanley Cohen, it is possible to recognize six kinds of vandalism.[22]

Acquisitive Vandalism. In this form of vandalism, property is damaged when money or goods are stolen. For example, vending machines or telephone coin boxes are opened because one wants the money inside these objects. A thief breaks the window or the door of a car in order to get to the stereo system or camera visible inside the car. The destructive behaviors in these cases are logical and understandable because destruction is necessary in order to reach the object wanted for manipulation.

Tactical Vandalism. In this variety, property is damaged in order to advance a certain objective, as a means to draw attention to a viewpoint, or to force a reaction. For example, prisoners destroy and burn furniture in order to make visible their living conditions and treatment in the prison, and to appeal to the general public for support. This is similar to a hunger strike in character; here one manipulates material objects instead of one's own body, as would be the case in a hunger strike. Vandalism of this nature is also understandable; there is an explainable reason for it.

Ideological Vandalism. This is similar to tactical vandalism, but the explicit purpose is to further an ideological cause. There is some specific ideological reason for doing so. For example, radical separatists such as the Irish Republican Army or the Basque separatists destroy offices associated with the authorities and governments that are opposed to them. In another instance, companies with business or commercial ties with South Africa may be vandalized by anti-apartheid groups.

Vindictive Vandalism. This refers to vandalism in which revenge is expressed by vandalizing objects that represent or are owned by a person or a corporate group, or a country that is seen as the source of damage one suffers. According to L. Taylor and P. Walton, one of the primary causes of the destruction in industrial sabotage is to express assertion of control by the workers. When interviewed, many of them explicitly stated that they had smashed things in order to increase their control and to show who was in charge.[23]

Play Vandalism. Here, vandalism is the same as play. A group of youngsters may, for example, play a game in which they compete as to who can break the most windows or paint the largest portion of a building wall.

Malicious Vandalism. This type of vandalism is a general expression of rage or frustration. This may be expressed indiscriminately, but often is directed against public buildings and facilities, such as subway stations, subway trains, public toilets, parks, and squares. This form of vandalism may be difficult to differentiate from play vandalism if one cannot ask the actor who has committed the act; when it is clear that rage or frustration is underlying the act, however, there is a function of satisfying an unful-filled need vicariously by expressing it through the manipulation of objects, and the result is seen as vandalism.

Vandalism is often described as motiveless, irrational, or senseless. But these six forms of vandalism can be easily explained in terms of stimulus seeking behavior. When one tries to reach for a certain object, anything that lies in one's way must be removed; otherwise it is impossible to get at it (acquisitive vandalism). When one is prevented from reaching for an object, either symbolic or materialistic, a substitute object nearby is manipulated (tactical vandalism). When the object that one wants to reach for is explicitly symbolic, such as the political ideologies of democracy, equality, freedom, or peace, objects that are seen to be associated with its prevention are demolished (ideological vandalism).

When one wants to damage an object that threatens one's well-being or existence and yet is unable to do so because that object is too strong, too remote physically, or legally impossible to attack, substitute objects that are seen to be associated with it are vandalized (vindictive vandalism). Certain material objects are chosen as targets of play, but because "wrong" objects are chosen, or because the play is expressed in a "wrong" way, the result is considered to be vandalism (play vandalism).

As an advanced primate species, *Homo sapiens sapiens* does experience rage and frustration, as do some other primate species, and it is inevitable that these feelings be expressed in action, because (1) we presumably have genetically acquired a very strong drive to manipulate, and (2) the expression of the manipulatory drive has an effect of reducing rage and frustration. But when a "wrong" object is chosen for expressing it, such as a subway train, or when the method of expressing it is "wrong," such as damaging a seat instead of cleaning it, the result is called vandalism (malicious vandalism).

Since vandalism is indeed a form of stimulus seeking behavior—albeit a disapproved form—to recognize the result of vandalism would give the vandal a sense of manipulation and control of the situation. According to Vernon L. Allen and David B. Greenberger, in their study of 120 vandalizing youngsters between 18 and 20 years of age, a number of responses indicated that a sense of control or mastery was an important factor in vandalism. For example, one boy felt a sense of accomplishment after breaking something. Another respondent saw the locker he had smashed in his high school, thinking proudly, "There's my little destruction to this brand new school," each time he passed it. This satisfaction lasted three years.[24]

There is also a series of studies in which vandalism is experimentally simulated, yielding similar results. For example, the feeling of success was greater after destructive behavior. It was more enjoyable to break complex structures than to break simple ones. Furthermore, to observe someone else's act of destruction has some effect upon one's perception of control, although not quite as much as when a person actually engages in the act personally.[25]

POWERLESSNESS AND VANDALISM

From our point of view, whether or not a certain behavior is vandalism is a matter of definition. When the physical act of destruction *per se* is seen, a "vandalistic" act may be less destructive than a socially accepted act. For example, breaking windows is not more destructive than shooting flying clay pigeons. The two forms of behaviors are strikingly similar to each other; in both cases, the objective is to derive satisfaction by destroying objects. Both are the "play" variant of stimulus seeking behavior, and the difference lies in the fact that one is socially disapproved and called vandalism, and the other is socially approved as a respectable sport to such an extent that it is one of the official sports included in the Olympic games.

One may argue that clay pigeons are made to be destroyed and since one is not damaging anyone else's property, clay shooting is not vandalism. But if we carry through this kind of argument, what about hunting? Is it less vandalistic to inflict pain and to kill animals for the sake of fun? I do not think so. Yet it is considered a sport rather than a kind of vandalism. Or consider environmental pollution. This is the most extreme form of damaging the environment for all living species as the most precious property. What about war? Pollution and war are not considered vandalism.

A conceivable reason for this distinction is the matter of power. When a certain act is committed by a person powerless in society, his or her act may be labeled as vandalism, but a similar or objectively even worse act by someone with power in society is not likely to be seen as vandalism. Indeed, among the six forms of vandalism as recognized by Stanley Cohen, tactical, ideological, vindictive, and malicious vandalisms can be interpreted as acts of those who are powerless in society.[26] When one cannot express one's manipulatory drive in the way one wants, one way to deal with it is to express it through a substitute channel, choosing another object and directing the manipulatory drive toward it instead.

This interpretation also makes sense in understanding deviant or criminal behaviors committed by children and youngsters. Juvenile deviant and criminal acts are to a significant extent manipulatory. In addition to play vandalism, shoplifting, stealing, and speeding are routine phenomena of juvenile delinquency. A common feature in all these behaviors is that the main focus is the manipulation of physical objects. When physical objects are manipulated in order to be destroyed, the result is called play vandalism. When objects are taken at will without paying for them in order to see that it is possible to manipulate them, the behavior is called shoplifting or stealing. When one manipulates a car excessively, the result is reckless driving or speeding.

The background of these behaviors can be found in the stage of life those who commit them are in. On the one hand, they are biologically mature, or almost mature, and physically strong. They may be even stronger than many adults. Therefore, they are likely to have a strong drive to manipulate. Yet, socially, they are almost totally powerless. They may be entirely dependent on their parents for living, and they may not be recognized as citizens by society; for example, they may not have the right to drink or to vote. Their views about life may be ignored by adults, and they may have difficulty influencing phenomena in society. Yet, because of their primatological background, they have genetically acquired the drive to manipulate, and when this drive is expressed in a way not approved by adults, their behaviors become deviant or criminal, depending on the applicability of the law.

This problem of vandalism and juvenile delinquency may to a considerable extent be due to the nature of the environment a youngster is surrounded by. When the environment is predominantly natural, such as forests, lakes, grasslands, and mountains, in most cases the application of the manipulatory drive to the environment is not disturbing to adults. To throw rocks into a lake, to break a tree branch, or to gather rocks at one place is not a problem to one's society unless it is very extensive or property is involved.

But when exactly the same behavior is directed against human-made objects by throwing rocks at a window, by breaking a bench in the park, or by piling up bricks in the middle of the street, these behaviors are stigmatized. When one is almost entirely surrounded by artificial objects as in the urban environment, then, there is little wonder that the problem of vandalism emerges. Biologically speaking, *Homo sapiens sapiens* has remained very much the same over at least 35,000 years or so as an advanced primate species, and its drive to manipulate has not changed. Its behavior based on this drive has not changed. Yet its environment has changed significantly, and, by merely having a different form of environment, the behavior of the human youth becomes undesirable and stigmatized.

Vandalism is almost exclusively play. This is beyond the stage of exploration, and its main focus of interest lies in seeing the object changed—that is, demolished and damaged. Some forms of creativity may be found in vandalism. For example, graffiti may be a source of novel effects, at least to the person who has done it.

SEXUAL DEVIANCE

The third category of the material we work on deals with human beings. This can be illustrated in terms of sexual deviance. When the behavior of sexual exploration is excessive, the result may become voyeurism, beyond the more condoned behavior of watching a striptease or pornographic magazine. Fetishism is a form of play in which one wants to interact with only a specific part of the body of another person, usually of the opposite sex, or its substitute, such as underwear or shoes. Creativity as expressed in sexual behavior may be seen in transvestism, mate-swapping, troilism (having sexual relations with or in the presence of more than one person), and many other forms of unusual behavior unfamiliar to most of us.[27]

In some cases, stimulus seeking behavior is carried out with violence, as in rape or sadism. When stimulus seeking behavior is directed toward a person not socially or culturally approved for that purpose, the result may be pedophilia (sexual contact with children), gerontosexuality (preference for older people as sexual partners), or incest. Sexual relations with a person who is socially defined to belong to a different racial or ethnic group may be, in a given culture, against the law, and in such a situation the behavior becomes a crime.

Stimulus seeking behavior dealing with sex may become more symbolic. Underwear or shoes in fetishism are symbolic in that these items stand for a human being. Transvestism is possible through identifying oneself with the opposite sex by means of symbolic thinking. In the case of exhibitionism, the result of exhibition, such as the way people respond to it, is seen as a successful manipulation of the environment. This, then, is a form of play.

In addition to sexual deviance, deviant manipulatory attempts involving human beings as objects also include child abuse, wife-beating, homicide, brawl, harassment, and mutilation. These behaviors may take place in bizarre forms as in sexual deviance. When one's own body becomes the object for manipulation, such phenomena as drug addiction, yoga, and suicide occur.

PSYCHOLOGICAL REACTANCE AND LEARNED HELPLESSNESS

As we have seen above, many forms of deviance and crime may be seen as expressions of the manipulatory drive that are judged to be unusual, excessive, or against the law. In these deviant or criminal behaviors, a person can indulge in stimulus seeking behavior. But in many situations, a person may not be able to manipulate objects that he or she wants to manipulate, or only to a very limited extent, or without satisfaction.

There are experimental studies dealing with this point. According to psychologist J. W. Brehm, we like to think that we have control over our behavior. One tends to assume that one is free to behave as one wishes, and this feeling of freedom is very important to the self. If one perceives that one's freedom is threatened or reduced, one tries to regain and reestablish it. This attempt is a form of motivation, which Brehm calls "psychological reactance." Brehm conducted a series of interesting experiments that corroborated his theory of psychological reactance.[28]

For example, in one experiment, he shows that a person's doing a favor for another may make that person feel that his or her freedom is threatened. As a result, in order to regain the perceived loss of freedom, he or she does not return the favor. This tends to happen when a person is made to feel that it is important to be free from obligation to another.[29] Experiments on psychological reactance, then, indirectly suggest the importance of being able to carry out one's manipulatory drive as one wishes.

In psychology, there is another category of studies in which freedom is more limited without the possibility for controlling the environment. To be in such a condition results in a phenomenon called "learned helplessness." Initially, this phenomenon attracted the attention of psychologists in their experiments on dogs. Dogs were presented with unpleasant stimuli, and they learned that they could not avoid them even when they responded in a variety of ways. Such an experience significantly interfered with the dogs' subsequent learning.[30] Similar experiments were conducted on other animals. For example, rats were exposed to stress in the form of unpredictable or uncontrollable electric shock. As a result, they showed a significant increase in ulcers.[31] In a study using wild rats held captive in an environment they could not control, it became clear that they lost incentive for living and displayed a phenomenon called "sudden death."[32]

For both human and nonhuman subjects, lack of control over the environment has a significant impact upon their cognitive, emotional, and motivational reactions when they are placed in stressful situations.[33] Among human subjects, there is an increase in depression, anxiety, and hostility among those who have developed learned helplessness.[34] It has been suggested that residents in nursing homes may develop feelings of helplessness and physical symptoms due to their inability to manipulate and control their environment.[35]

However, if we put it differently, if one believes that one can control the environment, even when that is not the case objectively speaking, one is less likely to develop these negative consequences. There are studies that suggest that one is less annoyed by noise when one believes that one can control the sources of noise.[36]

WORK SATISFACTION

In view of our strong drive to manipulate, the inability to manipulate is likely to be associated with a state of mind we wish to avoid. Or, to put it differently, the evolution of such a psychological response is conceivable along with the evolution of the strong manipulatory drive. In the literature of industrial sociology and labor relations, it is known that there is a positive association between control over one's work, on the one hand, and job satisfaction, on the other. The absence of (1) control over the pace of the work process, (2) control over the technical and social environment, or (3) freedom from hierarchical authority is associated with a high level of dissatisfaction with one's work.[37] Industrial sociologist Robert Brauner says: "It is possible to generalize . . . that the greater the degree of control that a worker has . . . the greater his job satisfaction."[38] There are enough data to support each of these three forms of dissatisfaction.

Absence of control over the pace of the work process is typically found in assembly-line work in the automobile industry. Assembly-line work is more disliked than any other major occupation, and the main reason for this is the lack of control over the pace of production.[39] Assembly-line workers want to change their work not because they want to get better pay, but because they simply want to get away from the line.[40] Over 90 percent of the workers in a study preferred to get a job away from the assembly line.[41] Quitting rates in this study were almost twice as high among men on the assembly line as among men off the line.[42] A study of 53 female operators on cannery conveyer belts also showed that all of them would prefer a different kind of work.[43]

Dissatisfaction is not due to the repetitiveness of the assembly-line work. Work at a machine in the automobile factory is just as repetitive, requiring as few motions and as little thought as assembly-line work. But work at a machine is preferred to assembly-line work because a worker can stop occasionally when he wants to. This freedom does not exist on the assembly line.[44]

In contrast to the workers on the line, miners have greater control over the pace of their work process, and, despite the lower prestige of their work,[45] they are more satisfied with their work. They can control the operation of their machines. The pace of machine operation and the area of the mine to work are determined by the workers themselves, and, even when the machine breaks down, the mine worker can make his own judgment as to what to do.[46] Although the surface workers have a higher status than the miners, they are less motivated than the miners because the speed at which the machines are operated and the procedures to be followed are prescribed by their superiors.[47]

Control over the technical and social environment is another way of expressing one's manipulatory drive, and whether or not this possibility exists in a job makes a difference in job satisfaction. Coal miners, steel mill workers, and railroad workers all can control their machines and materials, and all show a high degree of job satisfaction.[48]

In the case of white-collar workers, the environment they deal with is social, involving customers and clients. In this case, too, the realization that one has successfully manipulated the social environment is the source of satisfaction. For example, sales personnel derive such satisfaction from selling successfully.[49]

Freedom from hierarchical authority, a third source of job satisfaction, means that when a worker is not closely supervised he or she can determine how the work is to be done on the basis of his or her own judgment. The worker is the one who manipulates the work. But when the supervisor orders the worker how to do the work, the worker cannot look at him- or herself as the person who initiates the work. In the Aristotelian scheme of technology, one is merely the tool to be employed for doing the work.

Both railroad workers and truck drivers are not as closely supervised as, for example, factory workers, and the former are highly satisfied with their jobs.[50] The status of truck drivers is low in prestige,[51] yet they derive more satisfaction than do workers in all industrial occupations except railroad workers.[52] Here, again, the possibility for satisfying the manipulatory drive is more important to job satisfaction than prestige.

ALIENATION

Although it has been pointed out repeatedly that the concept of alienation is vague, ambiguous, or controversial,[53] in some studies of alienation the relationship between the individual and his or her inability to control the environment is the focus of interest. For example, according to Erich Fromm, an alienated person "does not experience himself as the center of his world, as the creator of his own acts—but his acts and their consequences have become his masters."[54] In the Aristotelian scheme, one cannot see oneself as the tool-user in this situation. Instead, the technique for using the tool ("his acts") and the result ("their consequences") are manipulating the self. These elements of technology become the tool-users, and the self as the tool-user is turned into the material to be worked on.

There are several factors that bring about alienation in the contemporary industrial world. For example, the Marxian version focuses attention on the system of production in capitalism in which the worker is forced to work in order to survive. The worker works not for him- or herself, but for the capitalist. Therefore, one does not belong to oneself but to another person when one works. The work is external and not a part of one's nature. One's labor becomes an object and takes on its own existence, which is independent and outside the self. It even stands opposed to oneself as an autonomous power.[55]

In this perspective, there are three changes in the way one looks at one's work in the Aristotelian scheme of technology. The worker cannot look at him- or herself as the tool-user any more. Instead, one sees oneself as a tool. The relationships between (1) the tool-user and the tool, (2) the tool-user and the result, and (3) the tool and the result have become disconnected.

Another version of alienation has been presented by Karl Mannheim. He points out that members of any particular generation have only a limited opportunity to have their impact on the affairs of their society.[56] Yet another version looks at a bureaucrat's insignificance in the huge bureaucratic system.[57] In these versions, the link between the tool-user and the result is seen to be weakened or disconnected.

AN ADAPTIVE ASPECT OF ALIENATION

In all of these three versions, then, an individual as the tool-user in the broadest sense of the term fails to see a causal link between him- or herself and the result of the work. The feeling of alienation in this situation results from the realization of one's own powerlessness, which has been pointed to as one of the characteristics of alienation.[58] The reason why a person experiences alienation in this form can be most logically explained in terms of the evolutionary perspective; that is, by having a strong drive to manipulate, a person strives to express his or her drive to manipulate the environment. By realizing that an attempt to manipulate has failed, a person may be in a position to experience alienation.

Alienation in this sense, then, is evolutionarily logical. When *Homo sapiens sapiens* has a very strong drive to manipulate, an individual may possibly try to express this drive in a wide variety of situations, involving both material and symbolic objects. Insofar as the expression of this drive is constructive, both for the individual and for the species, there is no evolutionary reason to discourage its genetic existence. But, in reality, there are two factors against unlimited exercise of this drive by every member of *Homo sapiens sapiens*.

First, we have been a species oriented toward the form of life based on groups, and, in a group, the strong manipulatory drive cannot be allowed to express itself in an unlimited fashion, because if everyone in the group tried to manipulate as much as he or she wanted, constant conflict, struggle, hostility, and killing would be inevitable. When a failure to manipulate results in alienation, an individual may experience depression, which has the effect of calming down the manipulatory attempt or preventing him or her from making another attempt to manipulate. To give up another attempt to manipulate is adaptive for group life.

Second, the manipulatory drive must be expressed in a constructive manner. Insofar as the amount of energy an organism can mobilize is limited, it is biologically harmful to utilize its energy in a useless or unproductive manner. When a member of *Homo sapiens sapiens* as a tool-user fails to see the relationship between his or her act of tool-using and the result of such act, the work is unproductive and useless; he or she is not accomplishing anything constructive. In this case, too, a built-in biological mechanism to prevent it is evolutionarily adaptive.

Thus, in order to prevent intragroup conflict, and also to discourage the waste of energy, the development of the psychological state called "alienation" is adaptive. By having the possibility for experiencing alienation, we avoid excessive and potentially fatal conflict, as well as the waste of energy, and survive better as a species. Alienation is adaptive in that it has the effect of calming down the individual. Through this built-in mechanism, a person slows down or stops his or her attempt to manipulate. At the same time, the alienated person realizes the futility of the attempt. He or she learns that the attempt to manipulate does not produce an effect. By means of alienation, the manipulatory attempt is calmed down, and the two undesirable consequences can be avoided.

Of course, this biological arrangement is not necessarily adaptive in all situations. Like almost any other form of evolutionary consequence, alienation has created conditions that are not adaptive for the species. For example, many occupations in industrial society are inherently prone to produce alienated workers, and this can hardly be considered adaptive. Furthermore, when a person has become alienated due to his or her work, he or she may lose all interest in life. The effect of alienation takes over those aspects of a person's life in which the active expression of the manipulatory drive is needed. But these phenomena occur because we are evolutionarily meant not for life in industrial society, but for that of hunter-gatherer society. The two adaptive functions of alienation are meaningful in the life for which we are biologically prepared, and that is life in a hunter-gatherer society.

SUICIDE

The phenomenon of suicide is complex, and, in reality, suicidal behavior is affected by such factors as culture, religion, personality, and unique incidents in a person's life. But certain cases of suicide, at least those of the type that sociologist Emile Durkheim called "egoistic," may be seen as a result of an expression of the manipulatory drive.

In Durkheim's terminology, egoistic suicide refers to the situation in which a person commits suicide due to excessive individualism without being integrated into the social network.[59] When one is relatively free from the constraints of social relations, one may be able to carry out what one wants or intends to do, and an extreme consequence of this may be suicide.

In terms of the perspective assumed in this book, a person has the genetically given drive to manipulate objects, which can be symbolic or materialistic. In normal situations, we are able to express the manipulatory drive in one way or another, because the total number of objects to manipulate in the environment is astronomical and practically infinite. This means that even when a person cannot manipulate the one specific object that he or she wants to manipulate, there is usually another object to manipulate, instead. For example, we often settle for the number-two item on the list when we cannot get number one, whether it be a grade in school, a spouse, a job, or a position in society. We often fail in our attempt to manipulate one specific object, but we usually get a similar one instead, even though it may not be as good as the one we first wanted. A student accepts a "B" on a test, and a man marries a woman even though she may not be "the most beautiful woman in the world," if I may borrow an old-fashioned expression from American popular culture. A new Ph.D. accepts a position at a small liberal arts college when a position at an Ivy League university is not available. This phenomenon constantly occurs in our daily lives.

A substitute object for manipulation may be totally different in nature. A man who has failed to become president of the country may find a new life in the world of business. A man may derive the satisfaction of power and money in the world of illegitimate business after failing in a business in a legitimate form. After a failure in ordinary life, one may devote oneself to a religious life as a monk or nun, manipulating esoteric, religious symbols. For practically everyone, then, no matter what happens in his or her life, there are almost unlimited possibilities for expressing his or her manipulatory drive. Certainly, members of some minority groups or of the lower class may find it difficult to do so in a respectable, legitimate manner, and this may explain their overrepresentation in the statistics of illegal activities.

However, what would happen if a person fails in his or her attempt to manipulate and subjectively cannot see any substitute object to manipulate instead? Other persons may clearly see a new life, a different career, or a new attempt in a different way after a failure, but the person in question may not recognize or accept a substitute object. At the same time, if, due to past learning, that person sees his or her own body as the only object that can be manipulated in a situation like this, suicide may result from such a perception of the situation. Here, one's own body becomes the object for expressing one's own manipulatory drive. This interpretation is, in a way, somewhat analogous to psychoanalyst Karl Menninger's view on suicide.[60] This can also be a conceivable explanation for certain suicide cases after a loss in economic status, a prolonged state of unemployment, or divorce.[61]

There are two categories of empirical data that can be interpreted to corroborate this view. First, it is well known that suicide rates went down when World War I broke out, and, after the war, they were higher during the Great Depression in the early 1930s and then went down again significantly during World War II. According to the data summarized by sociologist Jack P. Gibbs, which included Finland, Ireland, Japan, Norway, Sweden, and the United States, this pattern is unmistakably clear for all these countries except Ireland.[62]

It is certainly likely that the higher suicide rates during the Great Depression were due to the difficulty in perceiving an alternative object for manipulation for making a living. The drop in suicide rates during World Wars I and II may be interpreted as a result of the highly manipulatory power of war. By learning about or even personally seeing the destruction of cities, battleships, and fighters, as well as the killing of both soldiers and civilians, people were likely to experience vicariously the release of the manipulatory drive. But when the war was over, people could not do this any more, and it is worth noting that, in Japan, after losing World War II, officers of various ranks, and civilians most highly committed to winning the war, committed suicide.[63]

Second, another category of data deals with sexual differences. If we look at the primates as a whole, males are always more active in the manipulatory attempt.[64] In the case of humans, the influence of culture is significant, and variation can be considerable among both sexes. But in view of the fact that aggressiveness is dependent on circulating levels of the male sex hormone, testosterone,[65] it is possible to think that at least aggressive manipulatory attempts are more common among males than among females. A review of this issue indicates that boys are more aggressive than girls in their social play, even when they are only two to two-and-a-half years old. They have more hostile fantasies and engage in mock fighting. They threaten and attack each other in order to acquire dominant status.[66]

If this is actually the case in humans, it is logically conceivable that males are more likely to commit suicide than females because of their greater need to manipulate. With the more active drive to manipulate, males are more likely to attempt to manipulate and to do so to a greater extent than females. This in turn implies a greater risk for failure in the attempt. When a decision is made to commit suicide, males are also more likely to kill themselves actively and violently, resulting in successful suicide. Indeed, statistics consistently show this tendency regardless of culture.[67] Of course, actual suicide statistics may not truly reflect the reality of suicidal phenomena, and the tendency *per se* does not necessarily support the hypothesis presented here. Nevertheless, this is a plausible hypothesis for future research.

NOTES

1. Community Studies, Inc., *Teenage Vandalism* (Kansas City, Mo.: Community Studies Inc., n.d.), quoted in Marshall B. Clinard, *Sociology of Deviant Behavior*, rev. ed. (New York: Holt, Rinehart and Winston, 1963), p. 224.

2. Robert M. Lindner, "The Psychodynamics of Gambling," *The Annals of the American Academy of Political and Social Science*, 269 (1950), p. 97.

3. Chicago Area Project, *Juvenile Delinquency*, rev. ed. (Chicago: Institute for Juvenile Research and the Chicago Area Project, 1953), pp. 7–8.

4. Herbert J. Gans, *The Urban Villagers* (New York: Free Press, 1962), p. 65.

5. Frederic Thrasher, *The Gang* (Chicago: University of Chicago Press, 1927), p. 82.

6. George Grosser, "Juvenile Delinquency and Contemporary American Sex Roles" (unpublished Ph.D. dissertation, Harvard University, Department of Sociology, 1952).

7. Gerald C. Davison and John M. Neale, *Abnormal Psychology* (New York: John Wiley and Sons, 1974), p. 278.

8. M. Hotta, *Suri Gaisha: Moto Oyabun Shacho no Kiroku* (Pickpocket Company: The Memoir of a Former President) (Tokyo: Nihon Shuppan Senta, 1967), p. 160; T. Osatake, *Bakuchi to Suri no Kenkyu* (A Study on Gambling and Pickpocketing) (Tokyo: Shinsen Sha, 1980), p. 293.

9. Lindner, "The Psychodynamics of Gambling," p. 98.

10. Edmund Bergler, "Zur Psychologie des Hasardspielers" (On the Psychology of the Gambler), *Imago*, 22 (1936), pp. 409–41.

11. Thrasher, *The Gang*, p. 94.

12. Chicago Area Project, *Juvenile Delinquency*, p. 5.

13. Ibid., p. 5.

14. Ikuya Sato, "Play Theory of Delinquency: Toward a General Theory of 'Action,' " *Symbolic Interaction*, 11 (1988), pp. 191–212.

15. Frank E. Poirier, Anna Bellisari, and Linda Haines, "Functions of Primate Play Behavior," in *Social Play in Primates*, ed. Euclid O. Smith (New York: Academic Press, 1978), p. 156; Euclid O. Smith, "A Historical View on the Study of Play: Statement of the Problem," in Smith, ed., *Social Play in Primates*, p. 12.

16. Paul G. Cressy, *Taxi Dance Hall* (Montclair, N.J.: Patterson Smith, 1969), p. 84; Gans, *Urban Villagers*, pp. 64–73; William F. Whyte, *Street Corner Society* (Chicago: University of Chicago Press, 1955), pp. 35–51.

17. Poirier et al., "Functions of Primate Play Behavior," p. 155.

18. Albert K. Cohen and James F. Short, Jr., "Crime and Juvenile Delinquency," in *Contemporary Social Problems*, 3d ed., ed. Robert K. Merton and Robert Nisbet (New York: Harcourt Brace Jovanovich, 1971), p. 107.

19. P. Letkemann, *Crime as Work* (Englewood Cliffs, N.J.: Prentice-Hall, 1973); J. Lofland, *Deviance and Identity* (Englewood Cliffs, N.J.: Prentice-Hall, 1969), p. 73; N. Shover, "Burglary as an Occupation" (unpublished Ph.D. dissertation, University of Illinois, Department of Sociology, 1971).

20. J. Martin, *My Life in Crime* (New York: Harper and Brothers, 1952), p. 189.

21. Edwin H. Sutherland, *The Professional Thief* (New York: Holt, Rinehart and Winston, 1961), pp. 17–25.

22. Stanley Cohen, "Campaigning against Vandalism," in *Vandalism*, ed. C. Ward (London: Architectural Press, 1973).

23. L. Taylor and P. Walton, "Industrial Sabotage: Motives and Meanings," in *Images of Deviance*, ed. Stanley Cohen (London: Penguin Books, 1971).

24. Vernon L. Allen and David L. Greenberger, "Destruction and Perceived Control," in *Application of Personal Control*, ed. Andrew Baum and Jerome E. Singer (Hillsdale, N.J.: Erlbaum Associates, 1980), p. 87.

25. Ibid., pp. 95, 104, 101.

26. J. M. Martin, *Juvenile Vandalism: A Study of Its Nature and Prevention* (Springfield, Ill.: Charles C. Thomas, 1961).

27. J. L. McCary, *Human Sexuality* (New York: Van Nostrand Reinhold, 1967).

28. J. W. Brehm, *A Theory of Psychological Reactance* (New York: Academic Press, 1966).

29. J. W. Brehm and A. Cole, "Effect of a Favor Which Reduces Freedom," *Journal of Personality and Social Psychology*, 3 (1966), pp. 420–26.

30. M.E.P. Seligman and S. F. Maier, "Failure to Escape Traumatic Shock," *Journal of Experimental Psychology*, 74 (1967), pp. 1–9; J. B. Overmier and M.E.P. Seligman, "Effects of Inescapable Shock upon Subsequent Escape and Avoidance Responding," *Journal of Comparative and Physiological Psychology*, 63 (1967), pp. 28–33.

31. K. P. Price, "The Pathological Effects in Rats of Predictable and Unpredictable Shock," *Psychological Reports*, 30 (1972), pp. 416–26; J. M. Weiss, "Effects of Coping Responses on Stress," *Journal of Comparative and Physiological Psychology*, 65 (1968), pp. 251–60.

32. C. P. Richter, "On the Phenomenon of Sudden Death in Animals and Man," *Psychosomatic Medicine*, 19 (1957), pp. 191–98.

33. M.E.P. Seligman, *Helplessness* (San Francisco: W. H. Freeman, 1975).

34. R. J. Gatchel, P. B. Paulus, and C. W. Maples, "Learned Helplessness and Self-Reported Affect," *Journal of Abnormal Psychology*, 84 (1975), pp. 732–34; R. J. Gatchel, M. E. McKinney, and L. F. Koebernick, "Learned Helplessness, Depression, and Physiological Responding," *Psychophysiology*, 14 (1977), pp. 25–31.

35. R. Schulz and D. Aderman, "Effect of Residential Change on the Temporal Distance of Terminal Cancer Patients," *Omega: Journal of Death and Dying*, 4 (1973), pp. 157–62.

36. S. Cohen, "The Aftereffects of Stress on Human Performance and Social Be-havior: A Review of Research and Theory," *Psychological Bulletin*, 88 (1980), pp. 82–108; D. C. Glass and J. E. Singer, *Urban Stress: Experiments on Noise and Social Stressors* (New York: Academic Press, 1972).

37. Robert Blauner, "Work Satisfaction and Industrial Trends in Modern Society," in *Class, Status, and Power*, ed. R. Bendix and S. M. Lipset (New York: Free Press, 1966), p. 479.

38. Ibid., p. 479.

39. C. R. Walker and R. H. Guest, *Man on the Assembly Line* (Cambridge, Mass.: Harvard University Press, 1952), p. 62.

40. Ibid., p. 113.

41. Ibid., p. 110.

42. Ibid., pp. 120, 116–17.

43. Howard M. Bell, *Youth Tell Their Story* (Washington, D.C.: American Council on Education, 1938), p. 135.

44. Ely Chinoy, *Automobile Workers and the American Dream* (Garden City, N.Y.: Doubleday, 1955), p. 71.

45. Robert W. Hodge, Paul M. Seigel, and Peter H. Rossi, "Occupational Prestige in the United States, 1925–1963," *American Journal of Sociology*, 70 (1964), pp. 286–302.

46. A. W. Gouldner, *Patterns of Industrial Bureaucracy* (Glencoe, Ill.: Free Press, 1954), pp. 140–41.

47. Ibid.

48. E. A. Friedman and R. J. Havighurst, *The Meaning of Work and Retirement* (Chicago: University of Chicago Press, 1954), p. 176; John Spier, "Elements of Job Satisfaction in the Railroad Operating Crafts," unpublished paper, quoted in Brauner, "Work Satisfaction and Industrial Trends in Modern Society," p. 480; C. R. Walker, *Steeltown* (New York: Harper, 1950), p. 61.

49. Friedman and Havighurst, *The Meaning of Work and Retirement*, pp. 178, 106.

50. Robert Hoppock, *Job Satisfaction* (New York: Harper, 1935), p. 225; Joseph Shister and L. G. Reynolds, *Job Horizons: A Study of Job Satisfaction and Labor Mobility* (New York: Harper, 1949), pp. 13–14.

51. Hodge, Seigel, and Rossi, "Occupational Prestige in the United States, 1925–1963," pp. 286–302.

52. Hoppock, *Job Satisfaction*, p. 225.

53. F. Johnson, ed., *Alienation: Concept, Term, and Meanings* (New York: Seminar Press, 1973).

54. E. Fromm, *The Sane Society* (New York: Holt, 1955), p. 120.

55. Karl Marx, *Selected Writings in Sociology and Social Philosophy*, ed. T. Bottomore and M. Rubel (Harmondsworth, England: Penguin Books, 1956), pp. 171–78.

56. Karl Mannheim, "The Problem of Generations," *Psychoanalytic Review*, 57 (1970), pp. 383–84.

57. H. H. Gerth and C. W. Mills, *From Max Weber: Essays in Sociology* (New York: Oxford University Press, 1946), p. 50.

58. Melvin Seeman, "On the Meaning of Alienation," *American Sociological Review*, 24 (1959), p. 786.

59. Emile Durkheim, *Suicide* (Glencoe, Ill.: Free Press, 1951).

60. Karl Menninger, *Man against Himself* (New York: Harcourt, Brace and World, 1938).

61. Jack P. Gibbs, "Suicide," in *Contemporary Social Problems*, 3d ed., ed. Robert K. Merton and Robert Nisbet (New York: Harcourt Brace Jovanovich, 1971), p. 294; National Center for Health Statistics, *Suicide in the United States, 1950–1964* (Washington, D.C.: U.S. Government Printing Office, 1967), p. 7; Austin L. Porterfield, "Suicide and Crime in the Social Structure of an Urban Setting: Fort Worth, 1930–50," *American Sociological Review*, 17 (1952), pp. 341–49.

62. Gibbs, "Suicide," p. 299.

63. Michio Kitahara, *Children of the Sun: The Japanese and the Outside World* (New York: St. Martin's Press, 1989), pp. 80–81.

64. Jane Goodall, *The Chimpanzees of Gombe: Patterns of Behavior* (Cambridge, Mass.: Belknap Press, 1986), p. 339; Poirier, Bellisari, and Haines, "Functions of Primate Play Behavior," p. 155.

65. A. S. Chamove, H. Harlow, and G. Mitchell, "Sex Differences in the Infant-Directed Behavior of Pre-Adolescent Rhesus Monkeys," *Child Development*, 38 (1967), pp. 329–35; G. P. Sackett, "Exploratory Behavior of Rhesus Monkeys as a Function of Rearing Experiences and Sex," *Developmental Psychology*, 6 (1972), pp. 260–70.

66. Eleanor E. Maccoby and Carol N. Jacklin, *The Psychology of Sex Differences* (Stanford, Calif.: Stanford University Press, 1974).

67. Gibbs, "Suicide," p. 289.

4

CONFLICT IN SOCIETY

When one is faced with one's environment, it is conceivable that one tries to manipulate objects only when there is some possibility for manipulating them. It is a waste of energy to try to manipulate all conceivable objects indiscriminately. Hypothetically, it is plausible to assume that not only *Homo sapiens* but also a wide variety of animals have the ability to discriminate between objects that can be manipulated and objects that clearly cannot. This ability to discriminate between two kinds of objects is somewhat similar to the ability to distinguish between edible and inedible objects in the environment.

Our thought process dividing phenomena in the environment into two polar opposites such as edible/inedible, friend/foe, or safe/dangerous has attracted the attention of many, including C. G. Jung and Claude Lévi-Strauss, and this is probably one of the basic requirements in advanced primates, such as humans, for successful survival. Unlike lower level organisms, in which deterministic behavior toward the environment is dominant, we are more flexible in this regard, and the ability to categorize phenomena in the environment is prerequisite to the organism's attempts to deal with it successfully when there is no programmed pattern of behavior. Jane Goodall infers that this ability is also present among chimpanzees.[1] Eugene G. d'Aquili and Charles D. Laughlin suggest that this thought process takes place in the supramarginal and angular gyri and adjacent areas in the brain.[2]

The classification of objects into manipulatable and non-manipulatable categories is in part based on the level of technology; with efficient and powerful technology, objects that used to be considered not subject to manipulation become objects for manipulation. For example, the removal of a small island, landing on the moon, or the production of insulin by means of genetic engineering, in which animate or inanimate objects are involved, may be mentioned. A solution of a mathematical or philosophical or artistic problem, insofar as it is considered a successful solution, is an example involving an object of a symbolic nature. Yet, regardless of the level of technology, at any given level it is possible to divide objects in the environment in terms of whether or not the possibility for manipulation exists. For example, at the present state of technology, we do not think of changing the speed of the rotation of the earth, preventing an earthquake, or changing the direction of the wind.

To know that some objects are not subject to manipulation simplifies one's disposition toward them. One does not think of doing so, even when these objects are undesirable, unwanted, or dangerous. One has to live with them. This applies both to nonhuman situations such as natural catastrophes as well as to human situations such as inequality in a caste society. In both forms, people assume that it is futile and impossible to do anything about these things.

But when one sees some possibility for manipulation in such a situation and yet cannot manipulate or even reach for and try to manipulate a thing, a condition for tension and conflict is created. In history, for example, this is found in the incompatibility between social stratification on the one hand and the ideology of equality on the other.

THE IDEOLOGY OF DEMOCRACY AS A SOURCE OF CONFLICT

Western civilization has always been characterized by social stratification, and this is likely to remain the case as long as it exists. All complex civilizations throughout history are based on stratified societies. Aside from the explanations as to why this is so, stratification in society always accompanies a complex civilization.

Insofar as the members of the civilization accept that stratification is frozen and there is no possibility of changing it, they are unlikely to attack the government, the system, or those in power. Although there can be sporadic protests, one's intention is more likely to be that of expressing a grievance than seeking a change. After all, there is no sense of trying to change something that one thinks cannot be changed.

But when the perception of the environment changes through the introduction of a new ideology, people may begin to see more objects as within the range of manipulation. A new ideology offers a different perspective as to what can and cannot be manipulated. In the history of Western civilization, this new perspective is the philosophy of the Enlightenment.

The philosophy of the Enlightenment refers to a very comprehensive, large-scale system of thought involving many important philosophical concepts such as God, reason, nature, and humanity. Although some of the ideas attributed to the philosophy of the Enlightenment were found in ancient Greece and in the Roman Empire, the philosophy of the Enlightenment as a philosophical system emerged and developed in Europe during the seventeenth and eighteenth centuries.

It is extremely difficult to summarize such a large-scale philosophical system, but for our purpose here in this context, it is possible to point out three important features in it. First, power and authority were questioned. For example, philosophers such as Hume and Kant became skeptical toward the church and Christianity. Philosopher John Locke and statesman Thomas Jefferson questioned the power of the state. Economist Adam Smith also questioned the power of the state in controlling the economy and advocated the principle of laissez faire.

Second, a totally new concept of human beings emerged. It was assumed that human beings were good and reasonable. Humanity, love, and compassion were emphasized, and the mentally ill and slavery were scrutinized from a new perspective. Third, a belief in progress toward perfection was assumed in human history, as advanced by such men as A.R.J. Turgot and Marie Jean Antoine Condorcet.

One of the logical products of the philosophy of the Enlightenment is the ideology of democracy. By emphasizing the importance of the will of the people as a collective and also of the will of each individual, the ideology of democracy offered a new perspective in which the range of objects to be manipulated in society increased. If one wanted, one could now try to manipulate and change an object that hitherto had been either impossible to change or tabooed. For example, the right to vote became available as a tool in democracy, and so did the freedom of speech. By utilizing these means, one can think of manipulating one's society in one's own favor.

But the ideology of democracy offers a misleading picture of reality. Regardless of the political system of a given society, whether it be feudalistic, capitalistic, socialistic, or whatever, it is unrealistic to give the impression that one may be able to influence and change society. Without taking into account the fact that a society is made up of a wide variety of individuals of varied interests and objectives, the ideology of democracy

merely tells people that a democratic society is run by and for the people.
This promise gives people a false illusion of having more significant power
than they actually can exercise. Naturally, there is a risk of having a big
gap between what people assume they can manipulate and what they are
able to do in reality.

This point has been clearly indicated by Tocqueville in his well-known
book, *Democracy in America*, published in 1835 and 1840. He states that
"democratic institutions most successfully develop sentiments of envy in
the human heart. This is not because they provide the means for everybody
to rise to the level of everybody else but because these means are constantly
proving inadequate in the hands of those using them."[3] The problem of
democracy, as Tocqueville sees it, is that, on the one hand, democracy
promotes equality, and, on the other, it does not offer adequate means to
achieve equality. He further states: "This complete equality is always
slipping through the people's fingers at the moment when they think to
grasp it, fleeing . . . in an eternal flight. . . . They are excited by the chance
and irritated by the uncertainty of success."[4]

Here Tocqueville is describing the condition of people who think they
can reach for and manipulate objects, but yet find themselves unable to do
so in reality. Since this happens to *Homo sapiens sapiens*, which has such
a strong drive to manipulate, a strong reaction can naturally be anticipated.
In terms of the evolutionary perspective, failures to carry out the behavior
based on the manipulatory drive are unlikely to remain without significant
consequences. At least in the case of other drives such as sex, hunger, or
thirst, an unsatisfied condition is accompanied by a significant behavioral
change, and such a change is also likely to take place when the manipulatory drive is not properly satisfied.

In the milder form, a change is emotional. Tocqueville states: "the
excitement is followed by weariness and then by bitterness."[5] This has
also been pointed out by the German philosopher Max Scheler. According
to Scheler, the structure of democratic society itself creates feelings of
envy, anger, and hatred because, on the one hand, formal social equality
is publicly recognized, and, on the other, there are wide actual differences
in power, property, and education.[6] He uses the German word (borrowed
from the French) "*Ressentiment*" to describe this feeling. This word is
often translated into English as "resentment," but the two words are not
the same. *Ressentiment* refers to an attitude that arises from the repression
of envy: Superficially, one is nice to a person who is extremely lucky and
one hides one's true feelings of envy.[7]

The development of this complex feeling is understandable if we look
at the fundamental problems of democracy as an ideology. On the one

hand, there is the belief of equality, but, on the other, there is also the belief in achievement. Ironically, both are the products of the philosophy of the Enlightenment, and the whole problem here can be traced back to this inconsistency in the philosophy of the Enlightenment itself.

In addition to this sociological dilemma, a person experiences *Ressentiment* as an individual of *Homo sapiens sapiens*. His or her manipulatory drive is directed toward objects in the environment for manipulation. When one sees another person succeed while one's own manipulatory attempt fails, one is likely to experience the failure more intensely. Unless one identifies with the successful person vicariously and can have a pseudo-experience of his or her success, one is forced to realize, miserably, the inadequacy of one's manipulatory attempt. *Ressentiment* entails all these complex attitudes and feelings.

RELATIVE DEPRIVATION AND STATUS INCONGRUENCE

Two theoretical concepts that have been developed in sociology and social psychology, namely, relative deprivation and status incongruence, are relevant in explaining this response. Relative deprivation refers to the dissatisfaction one experiences when one cannot get what one thinks one is entitled to. For example, according to a study of American soldiers during World War II, despite the objective fact that soldiers with a high school education had better opportunities for advancement in the U.S. Army, they were not as satisfied with their status and jobs as were the less educated men. Similarly, Army Air Corps men were less satisfied with promotion opportunities than were men in the Military Police, although, objectively, opportunities for promotion were much better in the Army Air Corps than in the Military Police.[8]

In terms of our theoretical perspective, these findings can be explained as follows: Soldiers with a high school education perceive a higher rank and a higher salary as symbolic objects within the bounds of their manipulatory area. They assume that these objects can be reached for and held in their hands. Yet they discover that they cannot reach for them. As a result, they become frustrated. But soldiers without a high school education do not perceive the same higher rank and salary as objects within their manipulatory area. From the beginning, they do not assume that they can reach for and have them. Therefore, there is no reason to become dissatisfied with the fact that they do not get them. Soldiers in the Military Police, too, do not perceive a higher rank as an object to reach for and to grasp as commonly as do soldiers in the Army Air Corps. As a result, dissatisfaction is less frequent among soldiers in the Military Police than in the Army Air Corps. Dissatisfaction lies not in the objective but in the subjective domain, and depends in this case on how a soldier perceives objects around him as liable to manipulation. When a soldier assumes that he can reach for and have a higher rank within a reasonable amount of time and yet cannot have it, he becomes dissatisfied.

Another theoretical concept, status incongruence, is based on the assumption that an individual in society usually has more than one rank. For example, in comparison with others in the same society, a person may be ranked as to education, occupation, income, ethnicity, or race. People can be ranked on the basis of their occupations because different occupations usually carry a different amount of prestige. Similarly, it is common in the world that various ethnic groups are ranked formally or informally, legally or illegally, among the members of the same society. Level of education or income is even more objective and decisive in ranking people.

It is rather common that when a person occupies a high rank in one dimension, say, occupation, his or her income tends to be high, approximately corresponding to the amount of prestige that occupation carries. Furthermore, before such a person can occupy a prestigious occupation, he or she is likely to have a good education to begin with. For the same person, rank in one dimension tends to be consistent with rank in another dimension.

But this may not always be true. In some cases, a person occupies a high rank in one dimension but a low rank in another dimension. For example, a man may feel that his low position in a bureaucratic system does not reflect the amount of education he has received, or a doctor at a hospital may think he or she is underpaid. These are examples of status incongruence.[9] In these cases, one feels that a higher position or higher income is an object within one's reach but unattainable, exactly as in the examples of relative deprivation.

Both relative deprivation and status incongruence refer to very much the same phenomenon. It is possible to say that relative deprivation refers to the gap between two dimensions in status incongruence. Whichever perspective we choose, we are dealing with the same problem; a person thinks that a certain object can be reached for and had, yet in reality it does not turn out that way. The realization of this fact makes a person feel deprived or inconsistent in his or her statuses.

Feelings of envy, anger, or deprivation are common in any industrial society, which is dominated by all forms of bureaucracy and by a variety of conflicting interests. Insofar as people realize this as a fact of life and are more or less resigned to it, society remains calm on the surface, without large-scale conflict. Sporadic labor disputes and strikes occur, but, after mediation, grievances are calmed down, and people reluctantly return to work.

FRUSTRATION AND AGGRESSION

But what would happen if the expectation for reaching desirable objects is very high, or if the desired objects are very highly regarded, or if a person is forbidden to have the object after he or she has almost reached it? In these situations, logically, frustration and anger are much more intense than when expectation of reaching them is not high, when they are not highly regarded, or when a person is not close to them.

There is another theoretical perspective applicable to such a situation. Psychologist J. Dollard and his coworkers developed a so-called "frustration-aggression theory," and according to this theory, when a person is frustrated by not being able to reach a goal, frustration is changed into aggression.[10] This happens when the expected reward is delayed or not forthcoming at all. Aggression is assumed to be an innate response to frustration. This does not mean that frustration is always followed by aggression, nor that aggression is always caused by frustration.

The theory does not say that aggression is always directed toward the object that prevented the reward. Aside from the questions of under what conditions does frustration not lead to aggression, or under what conditions is aggression not directed against the perceived source of preventing the reward, certain conditions for attacking the preventing agent are conceivable. That is, (1) frustration is extremely intense, (2) the preventing agent is clearly recognizable, and (3) the retaliation or punishment from the preventing agent is not seen to be dangerous. Two kinds of studies on revolution can be quoted in order to illustrate this point of view.

In the 1920s and 1930s, three books on the most famous revolutions in the West appeared.[11] Among them, Crane Brinton's *The Anatomy of Revolution* is probably most widely known. In this book, the author examines the French Revolution, the Russian Revolution, the American War of Independence, and the English Revolution of 1640. Brinton recognizes seven characteristics common to all four events.

An Improving Economy. In all cases, the economy was improving prior to the revolution. The rebels were not starving nor were they oppressed. On the contrary, they were getting better economically.

Sharp Antagonism between the Social Classes. Strong resentment existed between the socially privileged aristocratic class and the class immediately below it. This class was only slightly below the aristocracy.

They were fairly close together to each other in status, yet one was clearly lower than the other.

Alienation of Intellectuals. Before and during the revolution, the regime was not supported by many intellectuals.

Inefficient Government. The government was not sensitive and responsive enough to changes in society, such as economic expansion, the rise of the new class with money, and a new technological development.

Loss of Confidence within the Ruling Class. The members of the ruling class began to doubt themselves and their class. Some of them deserted their class and joined the opposing group. Others lost the faith in their right to exist as a class.

A Financial Crisis. In Russia, a financial crisis was brought about by its involvement in World War I, and, in the other three cases, the financial administration of the state broke down.

Inefficient Use of Troops. When the revolt occurred, the government tried to suppress it by using troops, but they were no longer efficiently controlled by the government, and the revolt could not be stopped.[12]

Another study on revolution is that of James Davies.[13] In "Toward a Theory of Revolution," he examines economic conditions prior to Dorr's rebellion of 1842 in the United States, and those prior to revolutions in Russia and Egypt. In all three cases, the economy was good and society was not at all faced with a serious economic crisis. On the whole, people were getting along well, and their standard of living was constantly improving. Davies's conclusion is that "revolutions are most likely to occur when a prolonged period of objective economic and social development is followed by a short period of sharp reversal."[14] When this happens, people then subjectively feel that they have lost what they have gained to such an extent that they experience anger and frustration. The development of such feelings paves the way for a revolt.

In these two studies, then, we may recognize certain common elements that can be reinterpreted in terms of our approach. First, in both studies it is indicated that the economy is good and improving. Through time, the standard of living is constantly improving. In such a situation, it is logical to think that people want to have more objects both symbolically and materially, and when the economy is improving, they are more likely to get them sooner or later.

When this condition continues, they begin to assume that they can reach for more and more and actually continue to get more and more also in the future. Therefore, when such a trend is reversed by a sudden drop in their standard of living, people experience a shock. It is not only that they cannot reach for and get more and more objects, but also that they even cannot reach for and get objects that they could easily get in the past. The objects that could be easily manipulated earlier suddenly become objects that cannot be reached and manipulated. This experience is likely to create anger and frustration.

This reasoning may be corroborated by a well-known experiment in social psychology. In this experiment, children were allowed to play for a time with a set of attractive toys, such as a large dollhouse and a play pond with sailboats. These toys were all placed on one side of the room. After this enjoyable experience, they were faced with a totally new situation. A wire screen was drawn in the middle of the room as a partition, and this prevented the children from going to the part of the room where the attractive toys were placed. They could see the toys through the wire screen, but they could not play with them.

As a result, many of the children became furious and aggressive. For example, they kicked or shook the wire screen, and also threatened to hit the psychologist who was in the room with them. Some of them tried hard to circumvent the barrier by attempting to squeeze their hands through it, and protested and demanded that they be allowed to get at the toys.[15]

In the case of revolutionaries, there was an additional factor of anger and frustration. Brinton's study shows that there were two antagonistic classes in society that were very close to each other, yet were not the same. There was still a difference between the aristocratic class and the class immediately below it. Such a condition is most likely to create an intense feeling of anger, frustration, and resentment among the members of the lower class, because they are so close to the aristocracy but are forced to realize that they do not belong to it.

This point is illustrated by a study by E. G. Barber. According to her, the eighteenth-century French bourgeoisie was able to improve its social status in several ways. The available channels for raising status were: (1) the purchase of official positions, (2) acquisition of letters of nobility, (3) a military career, (4) an ecclesiastical career, and (5) the noble style of life. All of these channels were used to obtain the status of nobility. Only intellectual achievement and education raised status without the process of ennoblement. However, the nobles did not like to see these intruders trying to look and to behave like the nobles, and therefore closed these channels of upward mobility in order to prevent the bourgeoisie from becoming nobles. The bourgeoisie resented this humiliation and turned to revolution for a solution.[16]

Second, the agent that prevents the members of the lower class from reaching for and manipulating objects is clearly recognizable. It is the aristocracy or the privileged class that controls the government. In this sense, the target for releasing frustration is clear and unmistakable.

Third, Brinton's study shows that the government is unable to cope with the revolt effectively in order to suppress it. The intellectuals do not support the aristocracy and even desert it. The government is inefficient in coping with social change, and the members of the ruling class themselves come to distrust themselves. When troops are used to suppress the revolt, they turn out to be clumsy and inefficient. Such a condition of the ruling class and the government is most likely to encourage the tendency for revolt, because the members of the revolting class see not only little possibility for being suppressed and punished but also a greater possibility and hope for overthrowing the government for good. The possibility of getting hold of the government and the ruling class and overthrowing them is visualized. These objects are within the range of manipulation. This perception can mobilize a large number of people from the lower, antagonistic class to carry out the action of revolution.

In the studies by Brinton and by Davies, it is possible to infer the development of the feeling of relative deprivation among the members of the lower class. T. R. Gurr suggests that relative deprivation is the precondition for civil strife of any kind.[17] Yet it has been pointed out that not all forms of civil strife can be explained in terms of relative deprivation.[18] Such a finding may appear to disprove the interpretation of conflict in terms of relative deprivation. But another possibility is that relative deprivation is not a sufficient cause for people to act. They must certainly be dissatisfied with their lot by feeling relatively deprived, but they must also see some possibility in changing the situation by carrying out a revolt. It is conceivable that, no matter how people may feel dissatisfied, if they see no possibility for improving their lot, they are less likely to act in order to change. Relative deprivation is a necessary but not a sufficient condition for civil strife. At least the possibility of manipulating the establishment and the government as objects must be recognized by the members of the dissatisfied class.

A good example to be mentioned in this connection is the case of the Afro-Americans. Objectively speaking, compared with their earlier condition, the Afro-Americans gained significantly since 1954, when the Supreme Court of the United States declared that segregation of schools was unconstitutional. This was followed by a series of incidents such as the desegregation confrontation in Little Rock, Arkansas, in 1957, lunch counter sit-ins in the South beginning in 1960, mass gathering in Washington, D.C. in 1963, and so on.

Unlike the earlier sporadic protest actions in American history, the actions after 1954 have been more or less successful in improving the legal, political, and economic lot of the Afro-Americans as a whole. These improvements in turn made them realize the possibility for manipulating the establishment for their advantage, and the emergence of this perception is something new in the history of Afro-Americans.[19] Their assertive and militant protest behavior in the 1960s is a product of this new perception of the possibility for realizing their manipulatory drive.

PREJUDICE AND DISCRIMINATION

In this discussion of revolutions and large-scale civil conflicts, the key point is that individuals in the lower of the two classes feel deprived because they cannot have objects that they think they can have or are entitled to have, or, even worse, because they are prevented from having them any longer. Here, their perception of the objects in question is that they are within the bounds of manipulation. This becomes the source of anger and frustration.

Aggression is likely to follow when one cannot reach for and manipulate objects that one thinks are within one's reach for manipulation. It is conceivable that as long as one finds oneself in such a situation, regardless of the nature of the agent that prevents the manipulation, one is likely to respond similarly. The problem of racial and ethnic conflict, prejudice, discrimination, and persecution can be seen at least in part from this perspective.

One of the recognized factors in understanding the development of prejudice and discrimination is that of economic resources. If one can get what one thinks one is entitled to get, there is no possibility for developing anger and frustration. The object that one thinks is within one's manipulatory area is in hand; reality is in accordance with one's image of the world. But when reality turns out to be otherwise, and when one discovers oneself without the object that one thinks must clearly be in one's hands, anger and frustration are likely to develop.

In some cases, such anger and frustration may be turned toward oneself, in the form of blaming one's own lack of effort, knowledge, or skill in the attempt to get the desired object. Another possibility is to direct one's anger and frustration against the agent one perceives as having prevented one from getting the desired object. When this happens, it is logical to think that one hates and becomes hostile toward that agent. When possible, one may release one's anger and frustration directly against the agent. This is exactly he case with revolution.

In addition, as soon as one sees the possibility that an agent likely to prevent one from getting the desired object may intrude, one is likely to entertain fear, apprehension, and hatred even when one does not see any preventing action in reality. The perception of such a possibility is enough to create a feeling of this nature. This point has been expressed by D. Pierson as follows: "Race prejudice is usually acute in those situations in which members of a dominant group have come to fear that the members of a subordinate group are not keeping to a prescribed place of exclusion and discrimination but instead threaten effectively to claim the privileges and opportunities from which they have been excluded."[20]

Examples can be found, among others, in the history of the United States. When Chinese people began to immigrate to the United States, they were considered among "the most worthy of our newly adopted citizens," "our most orderly and industrious citizens," and "the best immigrants in California." They were spoken of as thrifty, sober, tractable, inoffensive, and law-abiding. They were described as showing "all-round ability" and an "adaptability beyond praise." These positive and favorable comments were made when their work was needed in California, and they were welcomed into domestic labor, cigar factories, and the shoe industry.

The attitude of the majority Americans changed drastically when they began to see the Chinese as competitors. In the elections of 1867, both political parties pledged themselves to enact legislation to protect Californians from Chinese competition. The Chinese began to receive a totally new description: "a distinct people," "unassimilable," "they keep to their own customs and laws," "do not settle in America," "carry back gold to their homes," "lower the plane of living," and "shut out white labor."

They were described as clannish, with dangerous secret societies, criminal, secretive in their actions, debased and servile, deceitful and vicious, and inferior from a mental and moral point of view. They were accused of smuggling opium and spreading the use of it, and Chinatowns were described as full of prostitution and gambling. The Chinese were "filthy and loathsome in their habits," "undesirable as workers and as residents of the country."[21]

This example clearly shows that competition for resources is enough to develop prejudice. There is a well-known experimental study that supports this hypothesis. Social psychologist Muzafer Sherif and his coworkers experimentally created a condition in which two groups of boys were made to develop a strong group consciousness. The members of each group felt that they belonged to a group that was competing against another group of boys. When they were faced with the situation in which two groups played a competitive game, each group was trying to get the limited resource—namely, winning the game. The competition in the game created intergroup hostility, resulting in name-calling and the refusal to associate with the members of the other group.[22]

In this experiment, the members of both groups saw the winning of the game as an object that they could reach for and have. This was perceived as an object within their reach. But the competing group was evidently trying also to reach for the same goal. They were, in effect, trying to prevent them from reaching the objective. The preventing agent was clearly recognizable, and, as a result, strong hostility against the other group developed.

To put this assumption differently, it is possible to say that people who are not in direct competition with a different group are less likely to be hostile or prejudiced toward them. It is known, for example, that retirement from active participation in a competitive society predisposes one to a greater tolerance toward the members of a different group.[23]

Also, if the members of a group are made not to compete against the members of another group and are told to cooperate together in order to achieve the same goal, hostility between the two groups is likely to be reduced. This assumption has long been known in the literature of sociology. For example, Georg Simmel saw that having an enemy country reduces conflict and hostility among various groups within one's own country.[24]

Experimentally, too, this hypothesis is supported. In the study of the two groups of boys we have just reviewed, this hypothesis was also tested. After the boys in both groups had developed strong hostility toward each other, they were told to cooperate toward a common goal, which was valuable to both groups. For example, they were asked to pull a truck up a hill to get it started again, as the truck was needed by both groups. After such experience, intergroup hostility was reduced, despite the fact that they did have a strongly hostile feeling toward each other before the cooperation.[25] This clearly shows that hostility is the result of competition, in which the other group is also trying to get the same resources. The members of the other group were disliked because they were seen as preventing the acquisition of the desired objects.

VIOLATION OF THE MATERIAL SELF

The problem of prejudice deals with competition over limited resources, which are accessible to others besides the individual in question and his or her own group. In this situation, one understands that the limited resources are in the public domain, so to speak; there is no guarantee or agreement that they exclusively belong to the individual and his or her own group. In this sense, the possible existence of a competitor is not completely ruled out, and one is aware of the danger of competition.

But if there is no such assumption, and if one's exclusive access to the limited resources is completely taken for granted, as soon as one sees another individual or group other than one's own trying to get the same resources, one is likely to be offended and to become angry. It is conceivable that the greater the intensity of the original belief, the greater the intensity of anger.

This can be illustrated by using William James's concept of the self mentioned earlier in Chapter 2. As you will recall, one of the three aspects of the empirical self is the material self, which includes one's own body, and the successive circles of things associated with it, such as one's clothes, one's family, one's home, and so on. By "the material self," anything that one thinks is one's own may be meant.[26]

Thus, if one realizes that a part of the material self is violated, disturbed, or invaded, one is likely to respond violently. One assumes that the material self belongs to oneself, and this conviction is likely to be very strong when that portion of the material self is close to the pure ego, or the self as the knower, such as one's own body, spouse, children, or highly cherished objects. As the psychological distance toward it increases, one may accept that others may also manipulate it, but they must be people acceptable to one, such as one's own family members, friends, members of the same ethnic, racial, religious, or socioeconomic group, and so on. In the literature of social psychology, there are experiments showing a person's tendency to assume the physical space immediately around the body as his or her own.[27]

The presence of an awareness similar to that of the material self can also be inferred among chimpanzees. For example, psychologist Wolfgang Köhler describes the behavior of a female chimpanzee named Tschego. When noisy chimpanzees came near her, she always grew angry, sprang up, stamped her foot, and struck out with her arms at the disturbers of her peace. If another chimpanzee came too close, she would seize one of his or her hands and give it a hard bite.[28]

Conceivably, the awareness of the material self may have evolved for the sake of the individual's survival.[29] Sociobiologist E. O. Wilson states that animals use aggression as a technique for gaining control over necessities, and most kinds of aggressive behavior among members of the same species are responsive to crowding in the environment.[30] In the study of animal behavior, if a fixed area is occupied by one or more individuals, that area is called a "home range."[31] When a part or all of the home range is defended, it is called a "territory."[32] If we follow this distinction, symbolically speaking, the material self covers all of the home range, and at least some of it is a territory.

Among our species, it is known that hunter-gatherers are generally aggressive in their defense of land where dependable food resources are available. According to Rada Dyson-Hudson and Eric A. Smith, hunter-gatherers defend those areas that are economically important to them. When food resources are scattered over a large area and not dependable over time, they usually do not defend their home ranges.[33]

If we use the concept of the material self, as discussed by William James, it is possible to understand that one may become hostile and angry when one feels one's neighborhood or school is "contaminated" by people who are ethnically or religiously not "entitled" to do so, or when one's country is "invaded" by immigrants or refugees. In these situations, one feels one's material self is violated, and hostility and aggression can easily be directed against the intruders.

SCAPEGOATING

When one feels that one cannot reach for and manipulate an object and yet cannot identify the agent that prevents one from reaching it, one is also likely to experience anger and frustration. Suppose an unemployed man cannot get a job. He knows he wants a job, but it does not seem to exist for him. He feels something is wrong with society, but he cannot quite understand why jobs are scarce. He may begin to believe that others are taking away jobs from him. For most people, it is difficult to clearly pinpoint the reason for a high unemployment rate, inflation, a high crime rate, and many other problems in society. Other forms of crisis may have clearly identifiable causes. For example, a country defeated in a war may look at the winning country as the cause for its chaos and problems. But it is impossible to attack and accuse the country that has won the war.

In these situations, an individual in society feels powerless at not being able to express his or her manipulatory drive; in the first situation, it is impossible to identify and attack and get rid of the agent that prevents one from having the desired object; in the second situation, it is impossible to attack and eliminate the agent that prevented one from getting the desired object.

In both situations, one may feel an urgent need to express the manipulatory drive in one way or another. To feel unable to express the manipulatory drive is likely to be a source of anger and frustration, and these feelings may become the motivating force for seeing the manipulatory drive realized in action. Scapegoating is one logical consequence.

According to psychologist Gurnek Bains, the successful rise of the Nazis to power through their blaming of the Jews may imply that the Nazis gave the Germans hope by attributing the crisis to causes that could be controlled easily. By pointing to the Jews as a highly visible minority group, the Germans thought they could deal with the problems in their society easily and successfully: The Jews were easily identifiable and small in number, and therefore it was easy for the Germans to arrest and exterminate them. All the problems could be solved by taking such manipulatory action.[34]

Even though most Germans may not have personally arrested, transported, and executed Jews as objects for manipulation, they could vicariously satisfy their manipulatory needs by learning about persecution and deportation. The ability to identify with another individual who can manipulate objects successfully and effectively and to think in terms of symbols among us has this kind of side effect, too. This is another unexpected consequence of human evolution.

Of course, the manipulatory drive can be released, even when the "cause" of frustration is clearly known, against an irrelevant object. In psychoanalysis, this is called "displacement," and, in ethology, both "displacement activity" and "redirected response" refer to the same form of response to frustration. Behavior of this nature has been observed in a variety of species including the starling, the blackbird, and the chimpanzee.[35] Experimental psychologists use the term "adjunctive behavior" to refer to apparently irrelevant behavior occurring in learning situations, such as grooming by a rat, and this is more or less the same kind of behavior.[36]

Another psychoanalytic concept, "sublimation," also deals with the expression of the manipulatory drive in an indirect manner. Irrelevant objects may be manipulated as in the case of engaging in a sport involving sporting tools and one's own body. Objects to be manipulated may be mostly or totally symbolic, as in the case of the sublimation of energy into arts and sciences. Other psychoanalytic reactions in response to frustration, such as reaction formation, intellectualization, and rationalization, can be considered as the expressions of the manipulatory drive in which symbolic objects are manipulated within one's mind.

SOCIAL CONTROL AND PUNISHMENT

When we realize that a society is made up of individuals, each of whom has his or her own drive to manipulate in his or her own way, the need for adjusting various attempts to manipulate is self-evident. This means actual prohibition and control of some forms of manipulation as well as attempts by certain individuals to prohibit and control them. In order to carry out such a plan, punishment is enforced upon those who violated the prohibition.

What sociologists call "total institutions" are arrangements in which the expression of the manipulatory drive is restricted by controlling freedom of movement and access to resources. By placing individuals in total institutions, those who are in a position to manipulate them as objects can enforce obedience, discipline, and conformity.

There are a variety of total institutions, and the extent to which one can limit another person's manipulatory activities varies from one to another. In the case of boarding schools or military barracks, there are opportunities to leave the total institution from time to time and to exercise the manipulation of the environment as much as most other individuals outside such a confine. Such a possibility is much reduced in other total institutions such as prisons and traditional mental hospitals in most countries. In sociologist Erving Goffman's words, "total institutions disrupt or defile precisely those actions that in civil society have the role of attesting to the actor and those in his presence that he has some command over his world."[37]

When a person's manipulation of the environment is limited to an extreme by coercion, we have such examples as slavery and concentration camps. In these situations, a person is not allowed to manipulate even his or her own body freely. In slavery, one does not own one's own body as the material self. It has been reported that in the Nazi concentration camp, defecation was strictly regulated, and prisoners who wanted to defecate had to obtain permission from the guard.[38]

Punishment may be intentionally directed against the material self of the person to be punished. Torture, tatooing, castration, amputation of hands, and other forms of bodily mutilation have been known in human history as methods of punishing individuals who expressed their manipulatory drive in ways unacceptable to those in power. In the case of the death penalty, a person loses his or her existence as an agent for manipulation.

NOTES

1. Jane Goodall, *The Chimpanzees of Gombe: Patterns of Behavior* (Cambridge, Mass.: Belknap Press, 1986), p. 588.

2. Eugene G. d'Aquili and Charles D. Laughlin, Jr., "The Neurobiology of Myth and Ritual," in *The Spectrum of Ritual: A Biogenetic Structural Analysis*, ed. Eugene G. d'Aquili, Charles D. Laughlin, Jr., and John McManus (New York: Columbia University Press, 1979), p. 163.

3. Alexis de Tocqueville, *Democracy in America*, trans. George Lawrence, ed. J. P. Mayer (New York: Harper and Row, 1988), p. 198.

4. Ibid.

5. Ibid.

6. Max Scheler, *Ressentiment*, trans. W. W. Holdheim, ed. L. A. Coser (Glencoe, Ill.: Free Press, 1961).

7. Fabio DaSilva, Anthony Blasi, and David Dees, *The Sociology of Music* (Notre Dame, Ind.: University of Notre Dame Press, 1984), pp. 160–61, n. 23.

8. Samuel E. Stouffer, E. A. Suchman, L. C. DeVinney, S. A. Star, and R. M. Williams, Jr., *The American Soldier*, vol. 1 (Princeton, N.J.: Princeton University Press, 1949).

9. E. Benoit-Smullyan, "Status Types and Status Interrelationships," *American Sociological Review*, 9 (1944), pp. 151–61; G. E. Lenski, "Status Crystallization: A Nonvertical Dimension of Social Status," *American Sociological Review*, 19 (1954), pp. 405–13.

10. J. Dollard, L. W. Doob, N. E. Miller, O. H. Mowrer, and R. R. Sears, *Frustration and Aggression* (New Haven, Conn.: Yale University Press, 1939); N. E. Miller, "The Frustration-Aggression Hypothesis," *Psychological Review*, 48 (1941), pp. 337–42.

11. C. Brinton, *The Anatomy of Revolution* (New York: Vintage, 1938; L. P. Edwards, *The Natural History of Revolution* (Chicago: University of Chicago Press, 1927); G. S. Pettee, *The Process of Revolution* (New York: Harper and Row, 1938).

12. Brinton, *The Anatomy of Revolution*.

13. James C. Davies, "Toward a Theory of Revolution," *American Sociological Review*, 27 (1962), pp. 5–19.

14. Ibid., p. 5.

15. R. G. Barker, T. Dembo, and K. Lewin, "Frustration and Regression," *University of Iowa Studies in Child Welfare*, 18, no. 1 (1941), pp. 1–314.

16. E. G. Barber, *The Bourgeoisie in Eighteenth Century France* (Princeton, N.J.: Princeton University Press, 1955), pp. 106–46.

17. Ted Robert Gurr, "Psychological Factors in Civil Violence," *World Politics*, 20 (1968), pp. 245–78.

18. D. Snyder and C. Tilly, "Hardship and Collective Violence in France, 1830 to 1960," *American Sociological Review*, 37 (1960), pp. 520–32.

19. Thomas F. Pettigrew, *Racially Separate or Together?* (New York: McGraw-Hill, 1971).

20. D. Pierson, "Race Prejudice as Revealed in the Study of Racial Situations," *International Social Science Bulletin*, 2 (1950), p. 473.

21. B. Schrieke, and O. Johannes, *Alien Americans: A Study of Race Relations* (New York: Viking Press, 1936), pp. 10–12.

22. M. Sherif, O. J. Harvey, B. J. White, W. R. Hood, and C. W. Sherif, *Intergroup Conflict and Cooperation: The Robbers Cave Experiment* (Norman, Okla.: University of Oklahoma, Institute of Group Relations, 1961).

23. Jeanne Watson, "Some Social and Psychological Situations Related to Change in Attitude," *Human Relations*, 3 (1950), p. 30.

24. Lewis A. Coser, *The Functions of Social Conflict* (Glencoe, Ill.: Free Press, 1956), chap. 5, proposition 9.

25. Sherif et al., *Intergroup Conflict and Cooperation*.

26. William James, *The Principles of Psychology*, vol. 1 (New York: Dover Publications, 1950), pp. 292–93.

27. Robert Sommer and Franklin D. Becker, "Territorial Defense and the Good Neighbor," *Journal of Personality and Social Psychology*, 11 (1969), pp. 85–92.

28. Wolfgang Köhler, *The Mentality of Apes* (London: Routledge and Kegan Paul, 1925), p. 248.

29. D. S. Lehrman, "Interaction between Internal and External Environments in the Regulation of the Reproductive Cycle of the Ring Dove," in *Sex and Behavior*, ed. F. A. Beach (New York: John Wiley and sons, 1965); K. Lorenz, *On Aggression* (New York: Bantam Books, 1966); N. Tinbergen, "Some Recent Studies of the Evolution of Sexual Behavior," in *Sex and Behavior*, ed. F. A. Beach (New York: John Wiley and Sons, 1965).

30. E. O. Wilson, *On Human Nature* (Cambridge, Mass.: Harvard University Press, 1978), p. 103.

31. W. H. Burt, "Territoriality and Home Range Concepts as Applied to Mammals," *Journal of Mammalogy*, 24 (1943), pp. 346–52.

32. G. K. Noble, "The Role of Dominance in the Social Life of Birds," *Auk*, 56 (1939), pp. 263–73.

33. Rada Dyson-Hudson and Eric A. Smith, "Human Territoriality: An Ecological Assessment," *American Anthropologist*, 80 (1978), pp. 21–41.

34. G. Bains, "Explanations and the Need for Control," in *Attribution Theory: Social and Functional Extensions*, ed. M. Hewstone (Oxford: Basil Blackwell, 1983), pp. 137–38.

35. Goodall, *The Chimpanzees of Gombe*, pp. 323–24; Niko Tinbergen, *Animal Behavior* (New York: Time-Life Books, 1965), p. 90.

36. J. L. Falk, "The Nature and Determinants of Adjunctive Behavior," *Physiology and Behavior*, 6 (1971), pp. 577–88.

37. Erving Goffman, *Asylums* (Garden City, N .Y.: Doubleday, 1961), p. 43.

38. Bruno Bettelheim, "Individual and Mass Behavior in Extreme Situations," *Journal of Abnormal and Social Psychology*, 38 (1943), p. 445.

5

HUMAN VARIATION

In the preceding chapters I have tried to demonstrate the existence of a strong drive to manipulate among primates, and have suggested that this drive is probably due to the evolutionary background shared by all primates. Since we are one of the primate species, it is logical and natural to think that this drive underlies our social and cultural behavior. This is one of the two basic assumptions of this book.

EVOLUTION OF VARIATION

The other basic assumption I would like to make in this book deals with variation. As a biological species, we are born, live, reproduce, and die in accordance with the principles of life. Whether we like it or not, our existence is programmed by DNA in chromosomes, which function as units for perpetuating genetic information.

Since every living organism dies, the genetic information that one particular organism carries can be preserved by another organism by means of reproduction. The most straightforward method of reproduction is mitosis, which can be compared with photocopying or carbon-copying. The genetic information is completely copied, and the copy is just like the original. This method is used in the binary fission of unicellular organisms. Mitosis is also the method found in asexual and vegetative reproduction in slightly more complex organisms such as the hydra.

There is another form of reproduction, in which the new organism is produced by two organisms. This is sexual reproduction. The new organism commonly develops out of the fusion of two separate cells, or gametes, such as an ovum and a sperm, derived from two separate organisms. Since this means the fusion of two sets of chromosomes, if the new organism is to maintain the same number of chromosomes for that particular species, it is necessary to halve the number of the chromosomes in the cells that function as gametes. This is achieved by a special division of the chromosomes called meiosis. If a species has $2n$ chromosomes, its gamete has n chromosomes, and these two states are called "diploid" and "haploid" phases, respectively.

The method of reproduction through these two phases has three significant implications in terms of the evolutionary perspective. First, during meiosis, homologous chromosomes form a pair, and, at this stage, sometimes an exchange of parts occurs between the pair, which is called "crossing over." This results in the mixing of genetic information.

Second, the diploid phase can store certain genetic information that may not overtly function because it is recessive. Latent genetic information of this kind may nevertheless become useful under different circumstances.[1] Third, the presence of the diploid phase reduces the risk of eliminating too quickly, genetically speaking, a recessive gene that may turn out to be adaptive under different conditions of life.[2] This is because a smaller proportion of recessive genes is expressed in the diploid phase, reducing the selection forces operating on and eliminating them.

Sexual reproduction occurs in a wide variety of species. For example, even paramecia conjugate and exchange genetic materials. Undoubtedly, the most obvious advantage of sexual reproduction is to increase the chances for a species' survival by being able to maintain variation within the species. This can be understood by comparing sexual and asexual reproduction.

In asexual reproduction, variation is possible only through mutation. However, mutation is known to occur very rarely. According to R. Sager and F. J. Ryan, spontaneous mutation rates are 2×10^{-7} for bacteria (*Escherichia coli*) and 1×10^{-6} for algae (*Chlamydomonas reinhardi*) per cell division.[3] In contrast, without mutation, sexual reproduction can have variations by having crossing over as well as by changing combinations of chromosomes.

This can be easily seen by looking at a very simple example. Suppose there are two alleles at two loci in a hypothetical genetic system, which might be expressed as A, a, B, b. From them, nine genotypes are possible: AABB, AABb, AAbb, AaBB, AaBb, Aabb, aaBB, aaBb, and aabb. In general, the total number of diploid genotypes from a species of n loci is 3^n. It is quite easy to see that the number of possible genotypes for almost any species relying on sexual reproduction, including ourselves, is astro-nomical.[4]

By means of sexual reproduction, then, it is possible to create much greater variation in a breeding population than in asexual reproduction. According to G. C. Williams, an asexual population consists largely of one or a few highly fit genotypes, but a sexual population is made up of a great variety of genotypes of lower average fitness.[5] Furthermore, genotype variety can function as a safeguard against environmental uncertainty, allowing the species to adjust better to a changed environment.

For this reason, sexual reproduction is to be considered a better solution than asexual reproduction to the basic problem of life. A species must maintain a certain distinct genetic basis in order to remain a species. It cannot have unlimited variation within the species; if that happened, speciation would inevitably occur, and a portion of the gene pool would disappear to form a different gene pool. In this sense, a species necessarily has variations within bounds. At the same time, the bounds cannot be too narrow. A species with too little variation cannot adjust to a significant environmental change, even when other species adjust to it successfully. A successful species has a proper balance between these two requirements in life. Therefore, sexual reproduction is a more satisfactory solution to this problem of life than asexual reproduction.

As an animal species having the sexual form of reproduction, we are characterized by variation. Obviously, variation can be visible in mor-phological and physiological differences among us. But, compared with other species, we are also characterized by significant psychological and cultural variations. These three forms of variation—namely, physical, psychological, and cultural variations—make us more successful in re-sponding to conditions in the environment than other species, but, at the same time, this very fact has ironically become a threat to our survival, as I shall indicate in later chapters.

GEOGRAPHICAL ENVIRONMENT AND VARIATION

It is easy to recognize that people are different from each other in a variety of ways, in terms not only of racial and ethnic differences, but also of differences within a breeding population. People are different in height, weight, skin color, hair color, the shape of the head, the relative length of arms and legs, the shape of the nose, eye color, the shape of the lips, the shape of the ear, and so on. Indeed, variations are impressive, especially in view of the fact that all living human beings belong to the same subspecies. Today, we all belong to *Homo sapiens sapiens*, and we can sexually exchange genetic materials between any two breeding populations. This fact of significant variation itself is a great advantage for us in survival.

By applying three principles derived from the study of mammals and birds, it has been pointed out that racial differences among us can be explained in part as adaptation to different physical environments.[6] These explanations are Gloger's rule,[7] Bergmann's rule,[8] and Allen's rule.[9]

Gloger's rule deals with the coloration of animals. In mammals and birds, races in warm and humid regions have more pigmentation than races of the same species in cooler and drier regions. This phenomenon can be seen in our differences in the color of skin, hair, and eyes more or less corresponding to the geographical regions of the world. Usually the variation in coloration among us is explained as an adaptation to the amount of solar radiation.

Bergmann's rule states that within the same species, races in the warmer part of the range of distribution tend to be smaller in size, and races in the cooler part tend to be larger. This phenomenon too is recognizable to some extent among us. Warm-blooded animals, such as human beings, must live by taking the factor of heat loss into account. That is, the smaller the surface of the body relative to the total volume of the body, the lower the loss of body heat, and vice versa. This means that a large person, who has a small surface area in relation to his or her total body mass, can withstand cold better than a small person, who has a large surface area in relation to his or her total body mass. Examples often given are the Scandinavians and southern Europeans.

Allen's rule points out that the shape of the protruding parts of the body can be seen as an adjustment to the temperature of the environment. Larger extremities facilitate heat loss; smaller protruding parts minimize it. Long arms and legs, a long and narrow head, long fingers, and a large nose perhaps make the life in hot climate easier by helping the organism release more body heat, and shorter arms and legs, a round head, short fingers, and a flat nose may help in reducing heat loss of the body. Africans and Eskimos are often mentioned as textbook examples in this connection.

Although there are recognizable exceptions to these rules, such as relatively light-skinned American Indians living near the equator, and large and stocky Polynesians in the South Seas, many human breeding populations can be seen in terms of these rules. If racial and ethnic differences are due at least in part to the differences in the geographical environment, to have a variety of physical forms is another advantage for us in survival. Each breeding population of *Homo sapiens sapiens* works as a genetic bank for the rest of us. If the climate on earth changes significantly in one way or another, those breeding populations having more adaptive genes suitable for that new condition may be able to survive better and at the same time supply the rest of us with the adaptive genetic materials. In this sense, considerable variations among us create a distinct advantage that we should be consciously aware of.

These three rules deal with phenotypic variations. There are additional variations in genotypes, which further increase the chances for survival. A well-known example is sickle-cell anemia. This disease is caused by a mutant recessive gene. When a person is homozygous for this recessive gene, he or she dies in childhood. But the heterozygous state is not lethal. Not only that, this genotype has a special immunity to malaria. The condition of one recessive sickle-cell gene makes the person malaria-resistant, and such a person can get on much better than those without a sickle-cell gene at all. The heterozygous individuals do not suffer from anemia, either.[10]

There are also speculations regarding the possible relationship between blood type and illness. For example, some researchers have claimed that the geographic variation of blood group frequencies is related to the prevalence of various infectious diseases in the historic or prehistoric past, such as plague, smallpox, and syphilis. But, according to geneticist Theodosius Dobzhansky and his coworkers, this matter is inconclusive.[11]

The possible relationship between racial variation and behavior is another topic that has been suggested and speculated on for a long time. In this case, the possibility of cultural influence upon behavior is obvious, and, for this reason, it is very difficult to derive any satisfactory answer. The factor of culture applies even during the growth of the fetus before birth through the way its mother eats, feels, and behaves, and much of this is affected by the culture she lives in.

Yet there are some studies that suggest that even when the influence of culture is taken for granted, racial differences may possibly explain some forms of behavioral differences. According to psychologist Daniel G. Freedman, there are significant average differences among various racial groups in locomotion, posture, muscular tone of various parts of the body, and emotional response, and Freedman thinks these differences cannot be explained in terms of cultural differences. For example, Chinese-American babies tend to be less changeable, less easily perturbed by noise and movement, better able to adjust to new stimuli and discomfort, and quicker to calm themselves in comparison with Euro-American babies.[12]

The difference between the two groups has been noted at a later stage of growth as well. In her study of nursery school children, Nova Green reports that between the ages of three and five, Chinese-American children are more individualistic and interact less with playmates. They show little intense emotional behavior. When they play, they make less noise, and their physical movements show more coordination.[13]

SEXUAL DIFFERENCES

In addition to racial and ethnic differences, we are of course characterized by sexual differences. The most obvious differences naturally deal with the anatomical and physiological ones directly related to the different roles males and females play in reproduction. But there are also differences in lifespan, propensity to illness, tolerance of stress, and chances for survival.[14]

However, the most significant difference between the sexes in terms of human evolution probably lies in the brain and nervous system. It is generally agreed that male brains are organized more asymmetrically for verbal and nonverbal functions than are female brains.[15] The evidence for this comes from the studies of patients with right or left cerebral damage. Male patients with a damaged left brain hemisphere perform poorly in verbal tests, as compared to nonverbal tests, but female patients are less likely to show such a result. Also, male patients with a damaged right brain hemisphere perform poorly in nonverbal tests compared to verbal tests, but again this is much less conspicuous among female patients. Some studies suggest that brain asymmetry is due to the male sex hormone, testosterone.[16]

As in the case of racial and ethnic differences, it is difficult to determine to what extent observed sexual differences in behavior are due to culture. However, studies of primates show sexual differences similar to the ones observed among many human cultures. In one study, a group of rhesus monkeys were separated from their mothers at birth and raised in social isolation for the first three years of their lives. They were then paired with infant rhesus monkeys of both sexes randomly. All the males raised in isolation directed significantly more hitting at their partners than the isolated females. The females raised in isolation showed nurturing behavior instead toward their partners.[17] Chimpanzees in their natural habitat are also known to show the same tendency. Although there is a considerable variation within each sex, males attack other chimpanzees significantly more frequently than females.[18]

Among human cultures, probably the Kung San of the Kalahari desert is one of the best examples in which the impact of culture upon observed sexual differences is minimal. In this culture, there is no difference in the way boys and girls are treated, and they are both raised with indulgence and permissiveness. Yet, even under such conditions, girls stay closer to home. In their play, boys imitate men, whereas girls imitate women. When they become a little older, sexual differences in behavior become more apparent. Girls care for smaller children and perform household tasks, and boys tend herds of domestic animals and protect gardens from intruding animals.

This difference is carried into adult life. Women gather nuts and other plant food, fetch water, and do household tasks, whereas men hunt. It is interesting to note that this occurs even in an egalitarian culture like theirs in which both sex roles are equally valued and can be reversed under certain circumstances. In terms of spatial mobility, men cover a wide space, but women's mobility is much more limited.[19]

GENETIC ABNORMALITIES

In addition to racial, ethnic, and sexual differences among us, there are also differences due to abnormal chromosomes. An individual of *Homo sapiens sapiens* usually has 46 chromosomes, but, among some individuals, there are more or less than 46 chromosomes. The X and Y chromosomes are so-called sex chromosomes, and, unlike the normal male (XY) or the normal female (XX), there are individuals, although numerically extremely small, who have one or more extra X or Y chromosomes or lack an X or Y chromosome. As a result, there are various forms of sex-chromosome abnormalities, such as the Klinefelter syndrome (XXY, XXYY, XXXY, XXXYY, XXXXY) and the Turner syndrome (XO, signifying the lack of one sex chromosome and only one X). These syndromes are associated with a defect in the sexual organs.

There are also sex chromosome abnormalities with the following combinations that are not characterized by a gonadal defect: XXX, XXXX, XXXXX, XYY, and XYYY.[20] Among them, no doubt the XYY combination has attracted the attention of the ordinary citizen in the street through the mass media. The XYY individuals are anatomically males (often called "super male"), characterized by tall stature, anti-social behavior, nodulo-cystic acne, and skeletal abnormalities. They rarely show any abnormality in sexual development but are reported to have a higher risk of exhibiting criminal behavior as compared with the normal XY males.[21] However, a study in Denmark suggests that they are no more aggressive than normal males, and that they are more likely to be on police record because of their tendency to be below average in intelligence.[22] Yet another well-known example of chromosome abnormality is Down's sydrome, which is associated with 47 chromosomes.

In addition to these chromosome abnormalities, there are other individuals who are characterized by abnormalities in their biochemical backgrounds, including enzymes, hormones, and neurotransmitters, resulting in abnormal health or behavior. Certainly, in terms of percentage, these individuals are quite insignificant, but that does not permit us to ignore them. Indeed, one of the most objectionable consequences brought about by the ideology of democracy is the cult of the majority. There is a false belief that what the majority is or wants is automatically good, right, and just. The implicit corollary of this belief is that those in the minority can be ignored. However, in view of the fact that our biological background is characterized by variation, this false ideology is most dangerous. No matter how small the number is, as long as certain individuals of *Homo sapiens sapiens* are different from the majority, their existence is in accordance with the biological principle of variation, and this fact must be accepted and the meaning of their existence must be consciously recognized.

PSYCHOLOGICAL VARIATION

Variations that can be described as "psychological" are observable in a wide variety of species. Even in honeybees and ants of the genera *Formica* and *Pogonomyrmex*, differences are known even within single castes. There are individuals that are unusually active, and they perform more than their share of lifetime work and incite others to work. But there are other colony members that are consistently sluggish. Their per capita productivity is only a small fraction of that of the diligent workers.[23]

Chimpanzees are characterized by even more variation. For example, differences are observable in terms of learning ability, intelligence and creativity, aggression, fear, motivation, throwing behavior, sexual behavior, mate preference, maternal attitude toward children, attitudes toward human beings, tantrum behavior among adults, and idiosyncrasy.[24] Jane Goodall remarks that "the marked differences in personality from one to another [among chimpanzees] are rivaled only by individual variation in our own species."[25]

Although it is conceivable that some forms of psychological variations such as the psychological differences between the sexes may be based on biology,[26] we can discuss psychological variations among human beings without necessarily connecting them to biology. For our purpose, it is sufficient to say that through evolution, the midbrain, cerebellum, medulla, and olfactory lobe, which control coordination and automatic activities of the body and personality, decreased in size, relative to the cerebrum, which is the apparatus for thinking.

This means that, as an advanced primate species, we have acquired a large cerebrum for thinking, which has become dominant in shaping our behavior, which differs from automatic, patterned behavior of less evolved species. The importance of thinking resulting from the large, convoluted surface area of the cerebrum in turn suggests a greater possibility for a variety of ways to think. Significant psychological variations, which always impress and surprise us, can be attributed to this biological basis.

Furthermore, there is also a neurological basis for psychological variation. According to Gerald Edelman, there are 26 major neuronal variation sites and levels, and, surprisingly enough, each of them has thousands to millions of possible variations! This naturally makes each brain unique. Even identical twins may be different from one another in terms of the size and shape of neurons, connection patterns, or amount of neurotransmitters. To put it simply, even identical twins who have the same genetic materials may be differently patterned neurologically.[27]

The psychological variation among us is almost a matter of commonsense, and we are well aware of the fact that people think and feel differently. There are numerous techniques for measuring various dimensions of personality differences such as the Minnesota Multiphasic Personality Inventory (MMPI), California Psychological Inventory (CPI), Taylor Manifest Anxiety Scale (MAS), Q-sort, Role Construct Repertory Test (RCRT), Draw a Person Test (DAP), semantic differentials, Thematic Apperception Test (TAT), Auditory Apperception Test (AAT), Rorschach, and Holtzman Inkblot Technique (HIT). This is enough to make us realize that the dimensions for variation, too, are numerous.

Casual observation of people also shows us that we feel and think quite differently from each other. The choice of marital partner is often a subject for entertainment. Some people like skinny partners while others marry overweight individuals. Height, the shape of the face, the nose, or the eyes, the length of the fingers and legs, and all conceivable physical features can be sexually attractive or unattractive. Even though the choice of one specific person as spouse is discouraged or proscribed due to a significant difference in age, height, social class, wealth of the family, or the racial, ethnic, or religious background, such marriages constantly take place.

Psychological variations are also observable in tastes, hobbies, eating habits, and various matters of likes and dislikes. Indeed, some individuals not only dislike certain situations, but also, when confronted by them, experience extremely intense fear in the form of phobias such as claustrophobia (fear of closed spaces), agoraphobia (fear of open spaces and of leaving the house), acrophobia (fear of heights), all of which are relatively well known. There are less familiar phobias too, such as hematophobia (fear of the sight of blood), hypnophobia (fear of sleep), pnigophobia (fear of choking), and taphephobia (fear of being buried alive). The fear of certain animals and insects exists among many of us, but, at the same time, there are individuals who love to have exactly those animals and insects as pets.

These phenomena look unusual enough to many of us, but there are even more extreme cases that are incomprehensible to most of us. We assume that food, sex, and life are, beyond question, things we like and strive for, and that it is unthinkable to give them up voluntarily. Yet there are individuals who do just that. Fasting is done for religious reasons or as protest. Sex is given up by some, often for religious reasons. There are individuals who undergo "sex change" operations, which in reality are not a change in sex at all but rather a very poor simulation of the anatomy of the opposite sex, without entailing the corresponding reproductive function.

According to the data collected by psychoanalyst Edith Weigert-Vowinkel, the Phrygians in Asia Minor practiced the cult of magna mater during the time of the Roman Empire, and followers of this cult who wanted to become priests ritually castrated themselves. In some variant forms, onlookers of orgiastic feasts of this religious cult also castrated themselves.[28] Self-mutilation of various parts of one's own body is known in various cultures, and in a bizarre form clinically called apotemnophilia ("amputation love") it is not uncommon for individuals with this affliction to cut off their own fingers or toes, or to damage a leg so severely as to necessitate its surgical removal.[29] It is totally unnecessary to elaborate on the fact that some individuals voluntarily give up their lives, sometimes through self-starvation.

As I mentioned earlier, chimpanzees show an amazingly wide variety of personality differences, but the practice of giving up or denying the expression of strong drives such as food, sex, and life is not found among them. Many animal species, including mice, cats, dogs, and chimpanzees, show neurosis-like symptoms either in captivity or in the wild, but probably only our own species develops psychosis. The psychological variation among us is indeed incredibly great.

CULTURAL VARIATION

Physical and psychological variations are observable among species besides us, and cultural variation regarding food habits, techniques of termite fishing, and cracking of oil-palm nuts is known even among wild chimpanzees.[30] However, there is little doubt that we show far greater variation in the cultural dimension than wild chimpanzees, and, further-more, cultural variation often means much more to us, in terms of adaptation and survival, than the other two forms of variation. For ex-ample, thanks to airconditioning or heating of buildings and efficient means of transportation, a person is able to live in other climatic regions of the world without being restricted to the specific region to which he or she is biologically adapted.

Similarly, the division of labor creates the possibility for a person to have an occupation in agreement with his or her interests and tastes. In a highly complex society it is possible to see very specialized occupations carried out by individuals psychologically suited to these positions. Or, to put it differently, because of the development of cultural complexity, psychologically varied individuals can find suitable niches in society, and this arrangement allows for the expression of human variation within the psychological dimension.

Cultural variation can be seen both intraculturally and interculturally. When a society is based on only one culture, intracultural differences can be found among social classes, occupational categories, different interest groups, and racial, ethnic, and religious groups. Differences of this nature are described by such words as subcultural or countercultural.

When cultural variation is considered in terms of differences among societies or nations, the issue is intercultural rather than intracultural. In reality, it is most likley that a citizen of a country is exposed to both intracultural and intercultural differences in his or her daily life. As in the case of psychological variation, we are constantly exposed to cultural variation directly or indirectly through the mass media, and there is no need to elaborate on this point. In many cases, the variation of culture is so significant that a person experiences "culture shock" when exposed to a totally different culture. Culture shock often deals with eating practices, sexual behavior, the way sexual difference is treated, and religious be-havior.

Like physical variations, cultural variations may be seen as adaptations to the physical environment. Geographer Ellsworth Huntington gives interesting examples in which many cultures can be logically explained in reference to environmental, geographical conditions.[31] In this sense, it is conceivable that a variety of cultures developed due to the variations in the physical and geographical environments. This, of course, does not mean that environment determines culture; the environment may limit the possibility for developing certain forms of culture but at the same time encourage the rise of one or more forms of culture as a response to one particular physical environment.

When we want to study ourselves and our behavior, then, we must take these three forms of variation into account: physical, psychological, and cultural variations. Human variation is interesting to observe but at the same time makes the study of human beings more difficult. Certainly, physical variation is found in many animal species, and this is subject matter for investigation in zoology and ecology. Psychological variation among other animals is studied by ethologists and psychologists, but variation is no doubt far greater among us. Cultural variation is almost exclusively limited to us, and, furthermore, in many situations this can be the most dominant form of variation in human behavior.

In this sense, the study of human behavior is handicapped by the fact that there are more variations than among other animals. It is not uncommon to come across a discussion comparing psychology (or sociology, or anthropology) with physics or chemistry in a book about psychology (or sociology, or anthropology), apologetically explaining why psychology (or sociology, or anthropology) is not as exact as physics or chemistry, and why prediction or theorizing is not as good as in physics or chemistry. The problem lies mostly in two of these three forms of variation—namely, psychological and cultural variations. Whether or not it is desirable or possible to predict or to theorize about human behavior is another matter; we must take these three forms of variation as given.

THE NATURE OF HUMAN VARIATION

The physical, psychological, and cultural variations among us have one characteristic common in all biological species: for each feature, we show a graded variation from one extreme to another. Relatively few individuals are on these two extreme sides, and most individuals are in the middle. One usually finds the bell-shaped distribution familiar in the study of statistics. Some examples are height, weight, aggressiveness, sociability, and attitude toward genetic engineering.

It is important to emphasize that a variation from one extreme to another is observable on a scale dealing with only one specific dimension when the issue is of an additive nature. For example, the weight of the human body is additive, in the sense that a person of 80 kg is 10 kg heavier than another person of 70 kg in the dimension of weight.

This is a relatively simple matter when we deal with physical variation, but with psychological and cultural variations the issue may involve a combination of two or more dimensions. For example, attitude toward communism may mean different things to different individuals. When people are asked about their attitude toward communism, one person may think of the difficulty of buying meat and of waiting in a queue for a long time, and another of Stalinism, and still another person may think of the sudden political changes in Eastern Europe, such as the dramatic downfall of Nicolae Ceausescu in 1989. Here, each respondent is thinking of different dimensions of communism, and, with each dimension, attitudes may vary from one extreme to another.

In many cases, the distribution of attitudes is symmetrical; there is likely to be about an equal number of extreme individuals on both sides of the distribution, and these people are about equally distant from the majority in the middle. But in other cases the distribution may be asymmetrical; the extreme individuals on one side may be much more distant from the majority than the extreme individuals on the other side. But even in these asymmetrical cases the general pattern is the same: there are extreme cases on both sides and the majority is in the middle.

There are certainly exceptions to this general pattern. It is possible that a nation can be sharply divided into two opposing camps, in which case most citizens show a strong attitude for or against a certain specific issue, leaving only a small portion of the population in the middle. It is also possible that a range of attitudes are shown by about the same percentage of the people in the nation, so that you get more or less a straight line instead of a bell-shaped curve. Yet, in reality, in most situations, the variations in the physical, psychological, and cultural dimensions are distributed in the bell-shaped form. Exactly for this reason, it has been possible to apply parametric statistics (statistics based on the assumption of the bell-shaped distribution of individual cases) to most data in medicine and the social and behavioral sciences, all of which deal with one or more of these three forms of human variation.

The fact that practically all forms of human variation are distributed in this manner has two important implications for us as a species in our physical, psychological, and cultural aspects. First, in almost any situation dealing with almost any issue, there is a small percentage of extreme individuals on both ends of the distribution who are physically, psychologically, or culturally significantly different from the majority in the middle. Second, as a corollary to this, in almost any situation dealing with almost any issue, it is possible to have a significant portion of the population as the majority, who more or less support, subscribe to, or agree with a certain middle-of-the-road solution to a problem.

But this state of affairs is not without a cost. To take the position of the majority at the same time means to ignore the positions of the extremes. The most ironic point here is that, on the one hand, the extreme positions, whether physical, psychological, or cultural, can potentially contribute to the survival of *Homo sapiens sapiens* by being a "bank" to preserve variation, but, on the other, they are likely to be ignored when we try to maintain our position without drifting too much toward one or the other of the two extreme positions. Thus, those positions that can potentially contribute most to the survival of ourselves under a different condition of life tend to be ignored.

This tendency can be seen in all three forms of variation. In the physical aspect of human variation, physically deviant individuals in a given breeding population have more difficulty reproducing themselves by attracting a member of the opposite sex less successfully. Extremely short or tall individuals, extremely obese or skinny individuals, extreme light or dark individuals, as judged in terms of the standard of that breeding population, are some examples of this. Similarly, in the psychological aspect of human variation, excessively radical or conservative individuals are much less likely to be supported by the majority.

In the case of cultural variation, it is possible to look at the problem intraculturally and interculturally. Within a given culture, there may be subcultures that are deviant from the culture of the majority, and these subcultures, such as the Oneida community or the Amish in American culture, tend to be ignored. Interculturally, in the contemporary world in which the deified ideology of "democracy" is a sacred belief in many cultures, countries based on racial stratification or religious dogmatism such as South Africa and Iran tend to be ignored or boycotted.

Of course, a country may be taken over by extremists, and insofar as such extremists can control and exercise their power over the country through oppression and persecution, their subculture is the official position of the country. If those in power can successfully indoctrinate the masses and make them believe in their rulers' extreme ideology, such ideology becomes the opinion of the majority and is no longer extreme. But if the rulers cannot convince the masses, they remain extremists.

Whether a given country is democratic, socialist, communist, aristocratic, theocratic, or whatever, there is always the same problem of the variation of individuals. If the ruling government represents the opinions of the majority, the extremists on both sides of the distribution are ignored. If the government is based on the ideology of the extremists, the majority in the middle and the extremists on the other side of the distribution are ignored. In both situations, some segments of the population are ignored. In general, it is possible to say that a government run by extremists is faced with more apparent or real opposition than the government based on the majority, and that such a country finds it more difficult to adapt to a changed condition in the world when the ideology of the extremists on the other end of the spectrum would adapt more easily to the new condition. This reasoning is likely to be applicable to a country like Rumania after the revolution of 1989, as compared with, for example, Hungary, although both countries were officially communist until 1989.

THE BIOLOGICAL REASON FOR CONFLICT

Like any other biological species, we are faced with the dilemma of how to survive as a species. On the one hand, we must have variation in order to be better able to adapt to changed environmental conditions, and, on the other, we must have a unity as a species. This dilemma is observable in our distribution curve. By having a small percentage of the population as extreme individuals on both sides of the distribution curve, variation is maintained, but, at the same time, having the majority of the population in the middle, unity is also maintained. Although this problem is found in all species, this is more serious and complicated for us because of our extensive psychological and cultural variations.

For this reason, conflict among us is logical and natural. The source of conflict lies in our background as a biological species. As long as we are characterized by variations in physical, psychological, and cultural features, it is inevitable to have conflict. Of course, conflict may not be manifest and visible all the time, but latently it is always present in the form of anger, dissatisfaction, and frustration among at least some segments of the population. To have these feelings means that we are characterized by variation as a biological species. So to speak, the source of conflict is built into our biological background as a species.

Conflict among us is additionally complicated by the fact that two or three forms of variation can jointly create conflict. For example, racism arises as a problem because we show a visible variation physically and also because people vary psychologically as to how we should evaluate racial variation. If all individuals of all racial groups were to evaluate racial differences uniformly in one way or another, conflict due to racism would not arise. Everyone could be racist and accept his or her racial background in accordance with the standardized ranking of the races. One accepts one's position in society in accordance with the official ranking. Or, alternatively, if all individuals of all racial groups totally ignored physical differences of peoples, naturally racism would not exist.

The evaluation of women until recently by both sexes was somewhat analogous to this situation. Both men and women took for granted that women were inferior to men and stupid. Since by and large both sexes believed this, there was no "sexism." Of course, in accordance with the fact of variation, there had always been a very small number of individuals of both sexes who thought and felt otherwise, but they were ignored for the reason we have indicated. Conflict over sexism emerged only after a significant portion of the individuals of both sexes in society began to think differently from the traditional evaluation of women. Any physical difference in age, height, weight, or function could become and indeed has become a source of conflict, and, in this case, too, physical variations among us are combined with psychological variations.

When the variations in the psychological and cultural aspects are combined, we have such examples as political or religious conflict, "class struggle," and persecution of a cultural minority. It is important to emphasize here that, although the way a person thinks may be to a certain extent influenced or conditioned by his or her culture, this is not a matter of complete determinism. Even in a totalitarian society, in which the sources of information are completely controlled by the state, there are always dissidents and rebels. This is easily understandable if we remember that we cannot get away from variation.

NOTES

1. G. G. Simpson, *The Major Features of Evolution* (New York: Harcourt, Brace and World, 1953).

2. H. J. Müller, "The Darwinian and Modern Conceptions of Natural Selection," *Proceeding of American Philosophical Society*, 93 (1949), pp. 459–70.

3. R. Sager and F. J. Ryan, *Cell Heredity: An Analysis of the Mechanisms of Heredity at the Cellular Level* (New York: John Wiley and Sons, 1961).

4. Peter Calow, *Life Cycles: An Evolutionary Approach to the Physiology of Reproduction, Development and Aging* (London: Chapman and Hall, 1978), pp. 76–77.

5. G. C. Williams, *Sex and Evolution* (Princeton, N.J.: Princeton University Press, 1975), pp. 151–54.

6. C. S. Coon, S. M. Garn, and J. B. Birdsell, *Races: A Study of Race Formation in Man* (Springfield, Ill.: Charles C. Thomas, 1950).

7. C. L. Gloger, *Das Abändern der Vögel durch Einfluss des Klimas* (Breslau: August Schulz, 1833).

8. C. Bergmann, "Über die Verhältnisse der Wärmeökonomie der Thiere zu ihrer Grösse," *Göttinger Studien*, 1 (1847), pp. 595–708.

9. J. E. Allen, "The Influence of Physical Conditions in the Genesis of Species," *Radical Review*, 1 (1877), pp. 108–40.

10. A. C. Allison, "Protein Afforded by Sickle-Cell Trait against Subtertian Malarial Infection," *British Medical Journal*, 1 (1954), pp. 290–94.

11. Theodosius Dobzhansky, Francisco Ayala, G. Ledyard Stebbins, and James W. Valentine, *Evolution* (San Francisco: W. H. Freeman, 1977), p. 161.

12. Daniel G. Freedman, *Human Infancy: An Evolutionary Perspective* (Hillsdale, N.J.: Lawrence Erlbaum, 1974).

13. Nova Green, "An Exploratory Study of Aggression and Spacing in Two Preschool Nurseries: Chinese-American and European-American" (unpublished M.A. thesis, University of Chicago, 1969).

14. Theresa Overfield, *Biologic Variation in Health and Illness: Race, Age, and Sex Differences* (Menlo Park, Calif.: Addison-Wesley, 1985), pp. 132–37.

15. J. McGlone, "Sex Differences in Human Brain Asymmetry: A Critical Survey," *Behavioral and Brain Sciences*, 3 (1980), pp. 215–63.

16. D. B. Hier and W. F. Crowley, "Spatial Ability in Androgen-Deficient Men," *New England Journal of Medicine*, 306 (1982), pp. 1202–5; G. Kotala, "Math Genius May Have Hormonal Basis," Science, 222 (1983), p. 1312.

17. A. S. Chamove, H. Harlow, and G. Mitchell, "Sex Differences in the Infant-Directed Behavior of Pre-Adolescent Rhesus Monkeys," *Child Development*, 38 (1967), pp. 329–35.

18. Jane Goodall, *The Chimpanzees of Gombe: Patterns of Behavior* (Cambridge, Mass.: Belknap Press, 1986), p. 341.

19. Patricia Draper, "Kung Women: Contrasts in Sexual Egalitarianism in Foraging and Sedentary Contexts," in *Toward an Anthropology of Women*, ed. Rayna R. Reiter (New York: Monthly Review Press, 1975), pp. 77–109.

20. Melvin M. Grumbach and Felix A. Conte, "Disorders of Sex Differentiation," in *Textbook of Endocrinology*, 6th ed., ed. Robert H. Williams (Philadelphia: W. B. Saunders, 1981), pp. 454–59, 476.

21. Ibid., p. 476.

22. H. A. Witkin et al., "Criminality in XYY and XXY Men," *Science*, 193 (1976), pp. 547–55.

23. Edward O. Wilson, "Man: From Sociobiology to Sociology," in *The Sociobiology Debate*, ed. Arthur L. Caplan (New York: Harper and Row, 1978), p. 231.

24. Goodall, *The Chimpanzees of Gombe*, pp. 25, 37, 39, 57, 68, 69, 72, 75, 144, 319, 336, 368, 436–37, 446, 557.

25. Ibid., p. 591.

26. Overfield, *Biologic Vairation in Health and Illness*, p. 139.

27. Gerald M. Edelman, *Neural Darwinism: The Theory of Neuronal Group Selection* (New York: Basic Books, 1987).

28. Edith Weigert-Vowinkel, "The Cult and Mythology of the Magna Mater from the Standpoint of Psychoanalysis," *Psychiatry*, 1 (1938), pp. 349, 351–53.

29. Armando R. Favazza, *Bodies under Siege: Self-Mutilation in Culture and Psychiatry* (Baltimore, Md.: Johns Hopkins University Press, 1987).

30. Goodall, *The Chimpanzees of Gombe*, pp. 248, 263, 561; W. C. McGrew, C.E.G. Tutin, and P. J. Baldwin, "Chimpanzees, Tools, and Termites: Cross-Cultural Comparisons of Senegal, Tanzania, and Rio Muni," *Man*, 14 (1979), pp. 185–214.

31. Ellsworth Huntington, *The Mainsprings of Civilization* (New York: John Wiley and Sons, 1945).

6

THE FALLACY OF CATEGORICAL
THINKING

The variations among us are gradual rather than abrupt, reflecting our biological background. This means that two or more individuals may appear to be rather similar to each other in regard to one specific aspect of human variation, such as skin color or aggressiveness. Also, we may get the impression that within a certain culture or subculture individuals may think and behave similarly.

To a certain extent, this assumption is useful and logical, because it is impossible to pay attention to all conceivable dimensions of physical, psychological, and cultural variations each individual has and decide to what extent a person is similar to or different from the majority in each of all these conceivable variations. As a result, we categorize people in one way or another and assume that all individuals in a certain category are the same. We economize our energy by doing so and make the world easier to comprehend and to deal with.

This approach is very practical when we are faced with the reality of the complex environment and have to decide which objects are to be manipulated, how, and when. But we also tend to forget that categorization of people is merely a device and that it distorts the reality of human variation. This mistake may be called the "fallacy of categorical thinking." This problem is known quite well among some circles in the scholarly world, especially among those who are interested in a field called general semantics.[1]

EXAMPLES OF CATEGORICAL THINKING

For example, in the history of Chinese philosophy, the issue of whether human nature is good or evil has aroused a continual controversy since the time of Mencius (371–289 B.C.?). Mencius said: "Man's nature is naturally good just as water naturally flows downward. There is no man without this good nature."[2] However, Hsün Tzu (active 298–238 B.C.) asserted that human nature is evil. By nature man is envious and hates others. Since man is by nature evil, man must be educated and disciplined, Hsün Tzu said.[3]

The futility of this debate lies in its assumption that people are similar to each other enough, psychologically, that they can be considered either good or evil as a whole. Although this assumption might be valid for some, it is certainly not applicable to extreme individuals. Despite their numerical insignificance, relatively speaking, the thought and behavior of these extreme individuals can be significant enough to make "good" human nature "evil," or "evil" human nature "good."

An additional problem is the assumption that human behavior can be "good" or "evil." In reality, human thought and behavior are extremely varied, and only a small portion of the total forms of human thought and behavior can ever be classified as either "good" or "evil." Of course, the concern of ancient Chinese philosophers was "human nature" and not human thought and behavior, but "human nature" is understandable objectively only through the study of thought and behavior. Additionally, a third problem is the relative nature of "goodness" and "evilness."

Another example is the Marxist explanation of ideology. According to the orthodox version of Marxism, human history is characterized by antagonism between the oppressing and the oppressed classes.[4] This assertion is logical, in that *Homo sapiens sapiens* varies psychologically, but it is not realistic enough because each class consists of individuals who think and behave differently. It may be reasonable to assume a tendency for thinking in a certain way according to the class one belongs to, but, at the same time, it is also necessary to realize that there are always individuals who think totally unlike others in the same class, or even like individuals belonging to another class. To equate class with ideology is a gross simplification that ignores human variation.

Marxism also insists that ideology is shaped by material conditions in society. Marx and Engels argued that when the means of production is owned by the state, the opposition to communism will disappear, because such ideas are the products of capitalism.[5] But it is far more in agreement with the fact of our biological background to assume, as, for example, sociologist Ralf Dahrendorf does,[6] that a revolution does not mark the end of opposition and conflict. No matter what kind of political system a people may have, it is more likely that there will always be some individuals who are dissatisfied with or against the system.

The persecution of dissidents, censorship of the mass media, jamming of short-wave broadcasts from the outside, restriction on foreign travel, and other restrictive measures that were enforced in communist countries until 1989 indicate that even when the means of production is owned by the state and no one is "exploited" by capitalists, people continue to show psychological variations, and some people are against communism. The mistake of communism lies in its very assumption of the lack of opposition in communist society; it ignores our background as a biological species. It is interesting to note that the assumption of Western, capitalist democracy is the opposite of this; it assumes that people think differently, and that political opposition is natural.

Of course, this does not necessarily mean that capitalism is better than communism, but at least its assumptions are slightly more realistic. Probably the absence of this assumption in the ideology of communism may explain the harsh power struggle between the Bolsheviks and the Mensheviks after the October revolution of 1917 and the era of Stalinist terror, as well as mainland China's "Great Proletarian Cultural Revolution" between 1966 and 1976 and the massacre at the Tiananmen Square in Beijing in 1989. When those in power are ideologically unable to realize the psychological variation of people, the only conceivable solution for them is to try to manipulate, control, and possibly exterminate dissidents. Variation is a concept alien to communists because of their political dogma.

In reality, this applies not only to communist revolutions, but to other forms of revolution as well. It is interesting and at the same time important to add in this context that internal power struggles among revolutionaries after the success of a revolution is a well-recognized phenomenon. By reviewing three important studies of revolution by L. P. Edwards, G. S. Pettee, and C. Brinton,[7] Jack A. Goldstone describes this phenomenon as follows: "Even where the revolutionary opposition to the old regime was once united, the collapse of the old regime eventually reveals the conflicts within the revolutionary opposition. After a brief euphoria over the fall of the old regime, divisions grow rapidly." Goldstone thinks this should be considered one of the ten lawlike empirical generalizations.[8] This suggests that revolutionaries are very prone to make the mistake of categorical thinking; they tend to look only at the difference between themselves and the establishment to be overthrown.

Surprise, disappointment, anger, and disillusion are common phenomena in political parties and organizations due to the mistake of categorical thinking. A political party is not made up of carbon-copied individuals having exactly the same political view and ideology. It merely means that the members of a party are relatively similar to each other compared with the members of another party. Each party is also characterized by the presence of two extreme sides and a majority in the middle, in accordance with our biological heritage.

CATEGORICAL THINKING AS A SOCIOLOGICAL PROBLEM

The fallacy of categorical thinking is also seen in the way criminality and deviance are conceptualized in sociology. There have been theories and hypotheses about how deviant or criminal individuals differ from conformists and noncriminals. In the history of criminology, we have seen theories in which some or most forms of criminal behavior are attributed to genetically influenced or determined factors such as body type.[9]

Another approach emphasizes the inadequate functioning of society as a system. According to sociologist Robert K. Merton, in the United States the means necessary for achieving the common goal of success are not equally available to all citizens. For example, the opportunity to get a good education may not be available to certain categories of individuals. Yet these people are supposed to aspire to such goals of success as high income, prestigeous occupation, and power. Naturally, there are individuals who realize that they cannot achieve the desired goal without the necessary means. As a result, some become professional thieves or white-collar criminals by accepting the goal but using an illegitimate means to reach it. Others ignore both the goals and means by becoming tramps, drug addicts, or dropouts of various kinds. Those who accept both the goals and the means are conformists.[10]

Yet another category of criminological theory emphasizes the factor of subculture within a culture. A certain subculture may be found in a section of a city, where behavior against the norms and the laws of the larger society is a fact accepted by the subculture. By conforming to the norms of the subculture, individuals within it become deviant and criminal as seen from the standpoint of the larger society.[11] In a version presented by criminologist Edwin Sutherland, criminals have their own subculture.[12]

These theories certainly deal with some aspects of deviance and criminality, and each of them can be supported to a certain extent by empirical data. But, at the same time, there is a serious defect in all of these theories.

That is, by classifying individuals in society in two or more categories, and by pointing out that individuals in a certain category are prone to criminality or deviance, it is assumed that everyone in such a category has an equal chance of becoming so. But this is far from true. Most individuals in the stigmatized category do not become deviant or criminal. This is one serious mistake of categorical thinking.

In fairness to sociologists and criminologists, it should be stated clearly that they are well aware of this problem; yet, in practice, they are forced to categorize people in society in terms of proneness to criminality and deviance. Indeed, there is not much choice for them. When criminologists must know something about deviance and criminality, it is necessary to classify people into two categories: those who have been judged to be criminal or deviant, and those who are not. This becomes the starting point for the study, and the basis of the study itself is already categorical in thinking.

The fallacy of categorical thinking also underlies prejudice and discrimination; evaluation of pupils and students in school; hiring, promotion, and firing of personnel; the choice of a marital partner; and many other decision-making situations in which we look at a specific individual in terms of his or her visible and tangible traits and backgrounds. In practice, we cannot live without categorizing individuals, but, by doing so, we are grossly distorting the reality of human variation. Insofar as we are consciously aware of making this mistake, the problem may not be so serious. But as some sociologists warn us, we can make a person "criminal" by treating him or her as a criminal and by labeling him or her as such.[13] Furthermore, once the categorization is done, it is not easy for us to change it, and serious distortions of reality result from this.

THE PROBLEM OF EQUALITY

Contemporary Western civilization is based on several fundamental assumptions, such as science, capitalism, and democracy. Both science and capitalism have a relatively long history in Western civilization; the ideology of democracy, however, in the contemporary sense of the term, is as recent as the eighteenth century.

Another assumption, one on the verge of becoming as important as these three assumptions, is the belief in equality. Although this belief may appear to be similar to the ideology of democracy, these two are distinctly different from each other. For example, it is well known that Thomas Jefferson, who signed the Declaration of Independence in 1776, owned slaves. Equality was applicable only to some of *Homo sapiens sapiens*, and, since the rise of the philosophy of the Enlightenment from which the ideology of democracy was derived, inequality of races as well as of the sexes was considered to be natural and was taken for granted for a long time.

Belief in equality is not yet truly established as an accepted ideology in Western civilization, probably because this began to be taken seriously only after World War II. The most decisive event was the Nazi Holocaust. This was followed by the civil rights movement in the United States in the 1960s, which in turn was followed by movements by various ignored and powerless groups, such as other racial and ethnic minorities, women, and homosexuals.

Compared with the other assumptions underlying Western civilization such as science, capitalism and democracy, equality as a belief may not have been critically evaluated and scrutinized. This is understandable if we look at the events that made us accept or think about accepting it as a necessary assumption to maintain Western civilization. The most common response to the Nazi Holocaust among most individuals, regardless of his or her background, was a sense of horror and shock. It was emotional rather than logical. In some cases, the basis for supporting equality was religious or intuitional. Whatever the reason, the belief in equality is based less on a long history of critical and evaluative scrutiny than are the beliefs in science, capitalism, or democracy.

This fact *per se* is in itself neither good nor bad. But one serious problem we have is that when we support a belief based on the intense feeling of emotion, we tend to react emotionally when there is any attempt to scrutinize and evaluate such an assumption. The intensity of emotion derived from the belief in turn becomes the intensity of emotion for supporting the belief. Any such attempt to look at the issue objectively becomes taboo. The belief in equality is now at such a point. There are several such emotionally charged issues of equality in the West, especially in the United States.

PROBLEMS WITH EQUALITY

As a result of the civil rights movement in the 1960s, Americans began to realize that racial minorities in the United States were disadvantaged in their attempts to succeed in American society, due to the poor quality of education they were likely to receive. Because they were educated poorly, they could not successfully complete higher education, which was and still is usually the requirement for a good occupation and a high income. These were all facts.

The ethical dilemma here for conscientious Americans was evident. On the one hand, one basic American value states that everyone is equal regardless of race, creed, religion, or sex; at the same time, however, some individuals are unjustly treated, in regard to the opportunity for education as well as in other ways. The most logical and natural solution anyone could conceive of was to equalize the quality of education, so that no one should be educated poorly. It was also assumed that by educating everyone equally, no one would be handicapped in competing for success in American society.

It was also logical and natural that researchers became interested in such an issue, one that was of national concern in the United States. But the results of their research surprised or disappointed many, and others became furious.

One of the first important findings appeared in 1966, in what is now commonly called the "Coleman Report," after sociologist James S. Coleman, who led the project. This study indicated that the unequal treatment of children in the early phase of education was not the primary cause of their unequal achievement at an advanced level of education. According to this study, the primary factors were beyond the control of the school. In determining achievement, the family background of the student was much more important than the quality of education he or she received. Children from the families with good incomes and occupational standing learned better than children from the families of lower standing in this regard. The quality of the school—the qualifications of the teachers, the teacher-student ratio, and the equipment for teaching—had little to do with the achievement of the students. The conclusion derived from the Coleman Report is that equal opportunity for a good education does not produce students who achieve at equal levels.[14]

In 1969 yet another work of importance and influence appeared, and this became the target of attack by emotionally upset Americans. The author was Arthur Jensen, and his article implied that, on the average, Afro-Americans genetically inherit a lower intelligence than that inherited by Americans of European background. This conclusion was based on data that suggest that, for most individuals, intelligence test scores are determined more by genetic factors than by environmental factors. In determining intelligence test scores, such favorable environmental stimuli as a good family background, good educational facilities, or well-qualified and competent teachers are less important than the genes one has biologically acquired.

For example, Jensen cites studies of identical twins, who tend to achieve similar IQ scores, even when they have been separated in infancy and raised in two different family environments. At the same time, another group of studies indicate that adopted children tend to make IQ scores similar to those of their natural mothers rather than to the scores of their adopted parents or siblings, even though the adopted children are in the same environment as their foster parents and siblings. If environment is more important than the genes, children from the families that encourage learning and intellectual development would do better, whether or not they have been adopted. But the findings suggest the contrary, that nature is more decisive than nurture in these twin studies, says Jensen.[15]

Another challenge to the optimistic belief in achieving equality appeared in 1972, when Christopher Jencks published a book entitled *Inequality*. Although this book differs somewhat from the works by Coleman and by Jensen in its focus of interest, its findings are consistent with these works, in the sense that Jencks emphasizes variation among people. Jencks and his coworkers show that there is no deterministic relationship between education and occupational status or income. The same educational qualifications may result in a wide variety of occupational and income variations. One's success in life may be based on one's effort as well as chance factors such as luck and accidents. Jencks indicates that even if everyone in the United States were to have the same level of education, unequal distribution of income would be reduced only slightly.

Jencks criticizes the policy of making education the same for everyone. Children are not the same, and therefore homogeneous schools are not suitable. Some children are motivated to study in a competitive environment, while others can do better without pressure, and yet another type of child learns better without an externally enforced standard of excellence.[16]

These three studies, by Coleman, Jensen, and Jencks, handle the problem of education in contemporary America in three different ways. But there is a feature common to all three. That is, all of these three studies call our attention to the phenomenon of human variation as one that we must possibly accept. It is extremely difficult to prove that there are differences in intelligence between one racial group and another. This would be possible only if we could show the complete arrangement of the heterocyclic amines of the DNA and how the assumed genetic influence on intelligence is expressed by a certain arrangement of these amines. Although this may come in time, this kind of information is totally unavailable now. But if we consider the fact that any biological species is characterized by variation, it is quite conceivable that, like any other biological attribute we have, intelligence also varies from one individual to another.

We know and accept variation among us in height, weight, or ability to run. We accept that various breeding populations of *Homo sapiens sapiens* are both genotypically and phenotypically different from each other. We even accept that some people are more intelligent than others insofar as this deals with individuals within the same racial group. It is strange that as soon as individuals of two or more racial groups are compared, highly emotionally charged accusations develop. Even to talk about such a possibility becomes taboo. This belief in equality is unsubstantiated and is based on the fallacy of categorical thinking. In this fallacy, it is assumed that all individuals of *Homo sapiens sapiens* have the same level of intelligence.

The problem lies in part in the vagueness or confusion of the word "equality." Sociologist Daniel Bell recognizes three possible meanings of equality." These are (1) equality of conditions, (2) equality of means, and (3) equality of outcomes. Equality of conditions mainly refers to public liberties, such as equality before the law, and political and civil rights. The underlying assumption for this kind of equality is equal treatment by a common standard. The equality of means refers to equality of opportunity. This deals with equal access to opportunities such as education, when education is the means for acquiring the competence necessary for achieving a desired position in society.

Equality of opportunity, according to Bell, is the overriding definition of equality in Western civilization, and, in the liberal tradition of the West, this assumption has never been seriously doubted. In contrast, the idea of the equality of outcomes is new in the Western tradition. But Bell points out that this form of equality can be realized only by restricting other individuals either in access to position or in the disposition of their achieved outcomes. The effort to reduce disparities of outcomes means the sacrifice of some people to others.[17]

If we look at this in reference to the other two forms of equality, we at once realize that, with equality of outcomes, both equality of conditions and equality of means are violated. Equality of conditions is violated because a person may be refused a desirable position on the grounds that he or she happens to belong to a racial or ethnic group overrepresented in that particular occupation. Clearly, this is a form of discrimination, and this is not equality before the law.

Equality of means is also violated because a person may be refused admission to a medical school on the grounds that a member of a minority group must be accepted instead. Here, the equality to access in the form of education is ignored and violated. To put it differently, this new form of equality emerged at the expense of the other two forms of equality established in the liberal tradition of Western civilization. The character of Western civilization is here forced to change by imposing this alien form of "equality."

Indeed, the belief in what Bell calls "equality of outcomes" is not based on traditional Western ideology at all. Rather, it is more logical to locate the origin of this belief in the primatological background of *Homo sapiens sapiens*. Everyone in society recognizes that society's resources, such as wealth, position, honor, and prestige, are limited. Some of these resources are material and others are symbolic. Both forms of objects are seen as objects for manipulation. Insofar as these material and symbolic objects can be reached for and manipulated, everyone in society is likely to attempt to do so in order to have these desired objects as his or her own objects.

Then how do people in a democratic society begin to assume that they can reach for and manipulate these desired objects? A conceivable answer lies in the two older forms of equality. By having maintained thus far the ideology of equality regarding conditions and means, people have assumed that outcomes would be equal as well. This is a great and in a sense tragic mistake, because people ignored—or, to put it more accurately, were not aware of—the fact of variation among individuals. The two older ideologies of equality made people assume that they were equal in all conceivable ways, including their abilities and outcomes.

Thus, the problem has two sources. On the one hand, the ideology of equality makes people think that desired objects are within everyone's reach. By making this assumption, people of varied ability and competence try to exercise their manipulatory drive. But, on the other hand, unable to manipulate the objects they assumed they could manipulate, they experience frustration and anger, exactly as in the situation before a revolution; they feel deprived in not being able to reach the goal they believed to be theirs. As in other cases of relative deprivation, anger and frustration emerge because people are made to think that they can get desired objects. In the case of this new idea of equality that emphasizes equal outcomes, people want to change the rules of the game by rejecting established forms of equality that deal with conditions and means, without considering the possibility that people may be different in ability and intelligence. Any suggestion and consideration of such a possibility is called "racist." This is a serious distortion of the ideology of equality.

THE PHILOSOPHICAL BASIS OF DEMOCRACY

In the contemporary world characterized by the development of communication networks of all kinds, probably one of the most used and abused words in politics is "democracy." As a word, it is a symbolic object that people want to have and to manipulate, despite the vastly different backgrounds peoples in the world have. This is a magic word, a key word in Western democracy, in Eastern Europe, and in third-world countries of both capitalist and socialist orientations.

In reality, "democracy" means different things to different political systems, and it is necessary to clarify this point. This can be done most effectively by scrutinizing the meanings of "liberalism," "democracy," and "liberal democracy."

In the history of Western social and political philosophy, "liberalism" can be found in the idea of social contract, as advanced, for example, by John Locke. Locke assumed that human beings were basically rational and reasonable. For this reason, Locke further assumed that human beings could live together harmoniously in society. Societies were to be formed by mutual consent because people were rational and logical enough to understand the virtue of forming and living in society. One would agree to give up one's individual power to defend one's own right against violators, and, in return for this, the community would have the duty to protect the rights of each citizen. The rights to be protected were life, liberty, and property.[18]

This philosophy of politics was taken over by Thomas Jefferson and was used as the justification for American independence; it was understood that when a government violated this contract, citizens had a right to dissolve it, reclaim power, and establish a new government. This was the reasoning behind the American rebellion, as expressed in the Declaration of Independence. For both John Locke and Thomas Jefferson, the rights one had were more important than the obligations one had toward the state.

The idea of social contract has been expressed in more than one way. In the case of Thomas Hobbes, no assumption was made that human beings were rational enough to live together harmoniously. Therefore, according to Hobbes, one had to surrender one's power to the sovereign power called the Leviathan, which in turn would protect the citizens.[19] Rousseau expressed his version of social contract by saying that the community to which citizens surrendered their rights was an expression of the general will of the citizens.[20] This means that to obey the general will means doing what one would want to do as an individual, according to Rousseau.

The common feature in these theories of social contract is that they all make assumptions about "human nature." Human beings are considered to be rational enough, or not rational enough, depending on the version one is investigating. But both are examples of categorical thinking, which

can be unrealistic enough to make any further logical deductions un-
tenable. If we remember the variations among *Homo sapiens sapiens* as a
biological species, it is far more realistic to say, in regard to the matter of
rationality, that some individuals are extremely rational, while others are
extremely irrational, and the majority is likely to be found in the middle.
Furthermore, an individual may not be rational or irrational consistently
through time. The same person can be both rational and irrational, depending on how and when he or she is studied. This factor of variation through
time is another characteristic we have.

The issue of living together to form a community can be a source of
variation, too. There are individuals who want to avoid contact with others
as much as possible, and there are those whose main concern is to be alone.
Such concepts as duty, obligation, or right that are part of the concept of
social contract can be yet other sources for variation. The concept of the
"general will" as formulated by Rousseau is an extreme example of
categorical thinking, in which it is assumed that a single "will" can be
derived from a population.

It is certainly conceivable that some individuals in the population may
find the "general will" as reflecting their own desires, but there are likely
to be many more individuals in the same population who disagree with the
"general will." According to the theory of social contract, these individuals
can or must see the "general will" as their own will, but they may find it
impossible to swallow in practice. To ignore variation and to enforce a
unitary pattern of thought and behavior is necessarily a source of friction
and conflict. It is quite ironic that the tradition of liberalism, which
presumably emphasizes the importance of each individual, has been made
compatible with such a solution.

The ideology of social contract persisted, more or less, in utilitarianism, and the importance of individuals was of prime concern to
both kinds of thought. The difference between social contract theory and
utilitarianism lies in the basis for forming the government. Unlike social
contract theory, utilitarianism sees the government as a means of utility
rather than as the result of a contract. According to Bentham, one by
nature avoids pain and seeks pleasure, and the role of the government is
to maximize these possibilities for its population. Thus, the fundamental
principle of utilitarianism, "the greatest happiness to the greatest number," emerged.[21]

But it is quite clear already in this expression that some individuals are
to be ignored. This principle, that one must give up certain things one likes
in order to fit into a utilitarian society, may be acceptable to most people;
but the happiness of deviant individuals may be sacrificed. This point has

been clearly pointed out by Tocqueville in his discussion of democracy in America. He says: "The moral authority of the majority is . . . founded on the principle that the interest of the greatest number should be preferred to that of those who are fewer."[22] Tocqueville also says: "What, then, is the advantage of democracy? The real advantage of democracy is not, as some have said, to favor the prosperity of all, but only to serve the well-being of the greatest number."[23]

Here, the main concern of liberalism—namely, the importance of the individual—is ignored in the name of democracy and in the name of the majority. Another irony is that utilitarianism, which has been derived from the liberalism of social contract theory, itself changed the character of liberalism and began to ignore the very principle of liberalism itself. By assuming that democracy, rule by the majority, is the best way to deal with the desire of each individual in society, a small number of individuals is ignored. This is a clear violation of, and stands in contradiction to, liberal principles.

John Stuart Mill was well aware of the problem created by utilitarianism. Because utilitarianism emphasizes that happiness is seeking pleasure and avoiding pain, it becomes logical to think that conforming to the opinions of the majority, even when an individual does not wish to do so, is the best solution, because that position is best in granting pleasure to the greatest number of people. Mill handled this problem by saying that there are different kinds of pleasures, and that some pleasures are better than others. In general, according to Mill, intellectual pleasures should have more weight than physical pleasures. He hoped that by means of education the quality of pleasure would improve, and that, when everyone would seek pleasures of a better quality, this problem would be solved.[24]

I personally like this view, but yet I must say that it ignores the possible variations of opinions within an educated population. Even in a country with a well-educated people, there is no guarantee that they will agree in evaluating various forms of pleasure. For example, even among intellectuals, a work of contemporary art or music may be evaluated quite differently. One artistic work of art can be a source of pleasure to one person, but the same work can be a source of pain to another.

In this sense, utilitarianism, just like social contract theory, is not compatible with the fact of human variation. In Bentham's version, some people are ignored in ways that contradict the principle of liberalism. At the same time, the tyranny of the majority may give deviant individuals pain instead of pleasure. Mill's approach does not guarantee a better solution because variation among us, in evaluating a variety of pleasure and pain, is not taken into account.

LIBERALISM AND DEMOCRACY

Although democracy and liberalism may appear to be the same thing, they are not. Democracy is a political system in which the government is run by the representation of the people. Admittedly, this is a matter of definition too, but a country can be "democratic" without being liberal, and vice versa.

For example, China is called "The People's Republic of China," suggesting that it is a "democratic" country, but China is not liberal, because there is no freedom for various political views nor freedom of speech. Yet it is possible to say that China is "democratic" because within the frame of communist ideology, people can vote and seek political office. Similar statements could have been made regarding the countries in Eastern Europe before 1989. The difference between Western democracy and communist democracy is that the range of choice in communist democracy is quite limited; yet, within such a limit, "democracy" can be practiced.

Conversely, it is also possible to have liberalism without democracy, as in the United States in the past, when women were not allowed to vote, or when citizens of African ancestry were not even recognized as human beings.

Liberal democracy, then, is a specific form of democracy in which liberalism is the basis for maintaining democracy. The welfare of each individual is taken for granted and placed over and above that of the community, at least in theory. The government must make it possible for each citizen to choose from a wide range of alternatives. Citizens are allowed to express different views and values. By maintaining the tradition of social contract theory and utilitarianism, liberal democracy assumes that each individual is rational enough to look at events in society carefully and can make sensible judgments. It is consequently assumed that by letting these citizens vote, a good government for the people can be formed. Since people are assumed to be rational, the different interests of different individuals can be adjusted to make society harmonious and peaceful.

Since liberal democracy contains liberalism as one of its elements, the problems associated with social contract theory and utilitarianism are also found in liberal democracy. It is quite unrealistic to assume that people are rational enough to make wise decisions in forming a government for themselves. Furthermore, since the principle of democracy is supported by the philosophy of utilitarianism, the tyranny of the majority is quite overwhelming to those who cannot accept it. As long as we are a biological species characterized by variations of all kinds, making everybody happy is indeed an impossible task.

The idea of liberal democracy has already acquired a large number of critiques and realistic evaluations. Edmund Burke did not believe that human beings are capable of choosing the best policy for themselves and also for the community.[25] Gaetano Mosca asserted that even when a country is officially democratic, it is not the people but an elite political class that rules the country. The ruling class consists of a small number of people, and, in order to justify their ruling, they use "myths" to make people accept their position in society. But, in reality, they never represent the people, according to Mosca.[26]

THE METAMORPHOSIS OF LIBERAL DEMOCRACY

When Arthur Bentley published his influential book entitled *The Process of Government* in 1908, the focus of interest shifted from the individual to the group. Along with this shift, the place of individual in liberal democracy disappeared. Bentley argued that politics is the struggle among interest groups to gain power to promote their interests and to achieve the goals of the group. As a result of conflict, struggle, and adjustment among the interest groups, the government is formed and run.[27]

Here, the key point is group versus group, and there is no place for the individual. An individual could possibly express his or her wish through the group, but, by taking this approach, the core principle of liberalism has been lost, because it is most likely that within the group people vary from one extreme to another. There are factions within a political party, and even an interest group is likely to be characterized by a bell-shaped distribution of attitudes and opinions. The extent of variation may be less than the variation in the whole population of the society in question, but even within the same party, extremists are likely to be ignored when the party makes its own decisions as to how to act against the other parties.

In a sociological study entitled *Voting*, published in 1954, Berelson, Lazarsfeld, and McPhee even argued that active participation by ordinary citizens in politics would do more harm than good, because this would routinely create a polarization of blocs and deadlocks, and it would be very difficult to carry out realistic politics under such conditions. In their view, it is much better to let more professionally minded politicians take care of politics.[28] By taking this position, then, the authors are abandoning one important aspect of liberalism, emphasis on the individual, which has been the major concern of both social contract theory and utilitarianism. I must admit that their conclusion is realistic, but I must also say that, for the sake of realistic politics, they ruthlessly scrap one very important part of traditional liberalism.

In this version of politics, the individual exists only as an element in a bloc, party, pressure group, or interest organization. One's voice can be heard only through such a collection of individuals. Although this may give us an illusion that, by participating in the activities of these organizations, everyone can influence politics, this may not necessarily be true. Since a group is made up of individuals, the most natural consequence is that opinions regarding a particular issue will vary from one extreme to another, as in any other situation of biological variation. The individuals whose opinions are similar to the majority opinion within the party or pressure group may be likely to have their voices heard, accepted, and taken up by the organization, but individuals with deviant opinions will most probably be ignored.

When the ideology of the organization already deviates from the majority of the society, in a realistic politics of compromise those individuals whose opinions are relatively close to the majority of the society have a better chance of advancing their views in order to avoid a deadlock. But the extreme individuals in the extreme organization, who in consequence of this deviate the most from the majority within society, have practically no chance in advancing their views. In this sense, group politics of this nature has a clear tendency to ignore extreme ideas on both sides of the distribution of opinions in society.

In this way, through time, the main focus of interest in the study of politics has shifted from the individual to the group. Despite the fact that liberal democracy finds its ideological origin in social contract theory and utilitarianism, in which the individual is of supreme importance, this assumption has been abandoned for the sake of the realities of politics. Politics as conceived by the theorists of contemporary liberal democracy is a continuous attempt to reach a compromise and a balance among a variety of political views, one that is carried out by professional politicians, or "elites," as they are called by those subscribing to this political theory, which is often called "pluralism."

THE REJECTION OF INDIVIDUALISM

Certainly, pluralism is a reality in Western democracy. Insofar as this form of politics is accepted as the most realistic solution in liberal democracy by the citizens themselves in this political system, they may not see any serious problem in continuing this way of life. But, in my opinion, it is very important to realize that pluralism functions at the expense of individuals significantly different from the majority. There is no place for deviant individuals in this distorted form of liberal democracy, and this goes against the biological fact of human variation.

It is quite natural that even in liberal democracy, which is supposed to be "open" and "free," there is a significant number of individuals who are dissatisfied, disillusioned, alienated, frustrated, or angry. It is also understandable that, as extremists, some of them turn to terrorism and violence. This all happens because variation among us is ignored and because the tyranny of the majority is enforced upon everyone. As an advanced primate species, every individual of *Homo sapiens sapiens* has a strong drive to manipulate, and when extremists find themselves unable or forbidden to express this drive, they understandably turn to terrorism and violence.

Contemporary liberal democracy exists by giving up the component of individuality in liberalism. Such a democracy functions for the majority at the expense of deviant individuals. Unfortunately, this may be the only realistic way to deal with the growth and complexity of contemporary industrial, as well as postindustrial, society. But precisely because of pluralism as it is conducted by political elites, there is always unresolved actualized or dormant dissatisfaction of the individual vis-à-vis society.

There are two reasons for this problem. First, in every individual of our species, there is the biological drive to manipulate. Having a view different from the majority opinion, a deviant individual may want to express his or her drive by all means possible, even when the result may mean criminal prosecution or life imprisonment. Second, despite the reality that deviant or extreme individuals are ignored in politics, liberal democratic societies continue to give the misleading impression that their citizens are running the country. For example, candidates for political office appeal to voters to vote for them, promoting the illusion that the voters will make political decisions through the politicians they vote for. But when professional politicians engage in pluralist politics, they must necessarily ignore the opinions of a significant portion of the voters who supported them. It is impossible for the voter to find a candidate who is in complete agreement with him or her on all conceivable issues in the world, and, furthermore, candidates are usually shrewd enough to avoid issues that are controversial or subject to debate. Voters are thus surprised or disappointed after the election because their politicians do not behave as they have expected.

This gap between assumed democracy by the people and actual democracy by the political elites can, theoretically, be dealt with in three conceivable ways. First, one possibility is to let the people know that they are not making political decisions in the way they believe, and that politics is run by political elites. Although this view is held by many alienated citizens who have realized their powerlessness in society, to state so explicitly may upset a significant portion of the general population, which naively believes in the idea of democracy by the people. This assumption of "democracy by the people" is so deeply rooted in the Western philosophy of liberal democracy that any politician in a Western democracy making such a statement runs the risk of being stigmatized.

Second, another approach is to give the impression that each citizen actually manipulates the outcome of politics. This is commonly seen in the ways citizens are urged to vote; they are made to feel that their votes can influence the course of politics. Political parties, too, make the same requests when they want voters to vote for them. Children are taught in school that they live in a democratic society and that it will be their duty to vote when they grow up. But an illusion is not a fact, and insofar as citizens are not intensely aware of the facts of democracy, society is relatively calm, peaceful, and free from open conflict. But as soon as they realize that politicians act against their wishes and their mandate, they suddenly become angry, and violent protest actions, riots, or terrorism may break out. In American politics, the issue of the Vietnam War, the Watergate scandal, and the Iran-Contra affair are good examples of this.

In a less dramatic form, in a democratic country there may be various offices both at the national and local levels in the public sector that are supposed to listen to citizens' grievances. Some examples are the offices for environmental matters, legal decisions, taxation, and consumer problems. These offices give citizens the illusion that their opinions are heard and that their ideas can change policy. Such possibilities may exist in some cases, but, unfortunately, in most cases they are, for the most part, mere window dressings for "democracy." Even in a country like Sweden, which is often considered to be one of the most democratic and enlightened countries in the capitalist West, bureaucrats in such offices simply lie, ignore issues, or remain silent whenever they are faced with problems they do not want to handle.[29]

Third, another possibility is to let citizens identify with the head of the state, such as the president or the prime minister, and thus satisfy their manipulatory needs vicariously. Here, if the leader is successful in letting a significant portion of the populace identify with him or her, the country is relatively free from dissent and conflict. A charismatic politician is one who is skillful at this. This approach is also common in actual politics. For example, in the United States, both the Democratic and Republican parties nominate the candidate who is likely to receive the most votes among the candidates within the party. Charisma, rather than political competence, is the key factor in choosing the party's presidential candidate.

Among the various theories of liberal democracy, regrettably, the theory of pluralism by political elites comes closest to the reality of Western democracy. Because this view was ignored when Bentley first presented it at the beginning of the twentieth century but later rediscovered in the 1930s and made popular in the 1960s, it is difficult to say that the theory stimulated the formation of contemporary liberal democracy. On the other hand, it is logical to say that this theory can clearly justify the present state of liberal democracy in the West as the logical and realistic solution to the problem of politics under both liberalism and democracy. But because traditional liberalism is distorted by the fact that it must ignore individual differences, contemporary liberal democracy as it is now in the West consistently bears within itself sources of alienation, frustration, disillusion, anger, violent protest, and terrorism.

NOTES

1. Alfred Korzybski, *Science and Sanity*, 4th ed. (Englewood, N.J.: Institute of General Semantics, 1980); Samuel I. Hayakawa, *Language in Thought and Action*, new rev. ed. (New York: Harcourt, Brace and World, 1964).

2. Wing-Tsit Chan, trans. and ed., *A Source Book in Chinese Philosophy* (Princeton, N.J.: Princeton University Press, 1963), p. 52.

3. Ibid., p. 128.

4. Karl Marx and Frederick Engels, "Manifesto of the Communist Party," in *Karl Marx and Frederick Engels: Collected Works*, Vol. 6 (London: Lawrence and Wishart, 1966), p. 495.

5. Ibid., p. 503.

6. Ralf Dahrendorf, *Class and Class Conflict in Industrial Society* (Stanford, Calif.: Stanford University Press, 1959).

7. L. P. Edwards, *The Natural History of Evolution* (Chicago: University of Chicago Press, 1927); G. S. Pettee, *The Process of Revolution* (New York: Harper and Row, 1938); C. Brinton, *The Anatomy of Revolution* (New York: Vintage, 1938).

8. Jack A. Goldstone, "The Comparative and Historical Study of Revolutions," *Annual Review of Sociology*, 8 (1982), p. 190.

9. Sheldon Glueck and Eleanor Glueck, *Physique and Delinquency* (New York: Harper and Row, 1956); C. Lombroso, *Crime: Its Causes and Remedies* (Boston: Little, Brown and Co., 1911); E. A. Hooton, *The American Criminal: An Anthropological Study* (Cambridge, Mass.: Harvard University Press, 1939).

10. Robert K. Merton, *Social Theory and Social Structure* (Glencoe, Ill.: Free Press, 1957), pp. 131–94.

11. Clifford R. Shaw and Henry D. McKay, *Social Factors in Juvenile Delinquency* (Washington, D.C.: Government Printing Office, 1931).

12. Alfred Cohen, Alfred Lindesmith, and Karl Schuessler, eds., *The Sutherland Papers* (Bloomington: Indiana University Press, 1956), pp. 5–43.

13. Howard S. Becker, ed., *The Other Side: Perspective on Deviance* (New York: Free Press, 1964).

14. James S. Coleman et al., *Equality of Educational Opportunity* (Washington, D.C.: Government Printing Office, 1966).

15. Arthur R. Jensen, "How Much Can We Boost IQ and Scholastic Achievement?" *Harvard Educational Review*, 28 (1969), pp. 1–123.

16. Christopher Jencks et al., *Inequality* (New York: Basic Books, 1972).

17. Daniel Bell, "The Public Household—on 'Fiscal Sociology' and the Liberal Society," *The Public Interest*, 37 (1974), pp. 56–57.

18. John Locke, *Second Treatise on Government* (Indianapolis, Ind.: Hackett Publications, 1980), chap. 9.

19. Thomas Hobbes, *Leviathan* (Harmondsworth: Penguin, 1981).

20. Jean Jacques Rousseau, *Social Contract* (Oxford: Oxford University Press, 1972).

21. Jeremy Bentham, *Fragment on Government* (Cambridge: Cambridge University Press, 1988); Jeremy Bentham, *Introduction to the Principles of Morals and Legislation* (London: Athlone Press, 1970).

22. Alexis de Tocqueville, *Democracy in America*, trans. George Lawrence, ed. J. P. Meyer (New York: Harper and Row, 1969), p. 247.

23. Ibid., p. 233.

24. John Stuart Mill, "Utilitarianism," in *Mill's Utilitarianism*, ed. James M. Smith and Ernest Sosa (Belmont, Calif.: Wadsworth Publications, 1969), pp. 37–42.

25. Louis I. Brevold and Ralph G. Ross, eds., *The Philosophy of Edmund Burke* (Ann Arbor: University of Michigan Press, 1960).

26. Gaetano Mosca, *The Ruling Class* (New York: McGraw-Hill, 1939).

27. Arthur Bentley, *The Process of Government: A Study of Social Pressures* (Chicago: University of Chicago Press, 1908).

28. Bernard Berelson, Paul Lazarsfeld, and William McPhee, *Voting* (Chicago: University of Chicago Press, 1954), chap. 14.

29. Letters from various Swedish authorities received by the author; copies are available upon request.

7

THE HISTORY OF CULTURAL EVOLUTION

The idea of evolution is one of the important products of the philosophy of the Enlightenment. Both biological and cultural evolutionism implicitly or explicitly assumes that, over time, there is progress and improvement. Evolution is something good and desirable. For philosophers such as Herbert Spencer, who believed in the principle of laissez faire, evolution was a cosmic and universal law: we could observe continued progress and improvement by merely sitting back and watching the world and the universe.

Others such as Auguste Comte and Karl Marx felt the necessity for the active involvement of human beings in the process of evolution, in the form of a new religion (Comte) or revolution (Marx). Despite these differences, there is one consistent and common feature in the ideology of evolutionism. Evolution is good and desirable, and it is something that we should look forward to with great expectation and hope. Evolution is a blessing to everyone. In a contemporary version, this optimism is expressed in a more practical manner, such as the idea of a continual increase of energy available per capita.[1]

More than 300 years after the emergence of the philosophy of the Enlightenment (*De jure belli ac pacis* by Hugo Grotius was published in 1625), we are still almost completely immersed in this philosophy, and it is very difficult to look at ourselves, our history, and our future in a way free from it. After all, our current education is basically a product of the philosophy of the Enlightenment, and the very foundation of industrial society is also based on it, and it does not make any difference whether or not we are members of the capitalist West. We take the philosophy of the Enlightenment and its products for granted as natural and proper. Or, to put it more accurately, we do not think about it because this way of thinking completely dominates us.

But is this right? Can we afford to be so optimistic? Is it not necessary to look at our past and present carefully and to reconsider what *Homo sapiens* has experienced in its very short existence as a biological species? In this chapter, I would like to take up these questions. In Chapters 1 through 4, I discussed the phenomenon of stimulus seeking behavior and its evolution among primates. On this basis, I assumed the corresponding evolution of the manipulatory drive from lower level primates to *Homo sapiens*. In Chapters 5 and 6, I discussed another important phenomenon of variation. I assumed two axiomatic postulates in trying to understand human behavior, history, culture, and civilization: (1) stimulus seeking behavior with the underlying manipulatory drive, and (2) variation. On the basis of this assumption, I shall now look at the history of *Homo sapiens* as a biological species in three phases: (1) hunter-gatherer society, (2) agricultural society, and (3) industrial society.

Before going into details, however, it is necessary to add a few words about this approach. The cultural evolution of *Homo sapiens* is an extremely complex matter, and it is impossible to describe it without simplification. To conceive of it in terms of three phases is clearly a simplification, but, at the same time, it is a realistic solution to the impossible task of looking at our cultural evolution. The only justification for using this conceptual device lies in the fact that human life changed significantly after the neolithic revolution (discussed later in this chapter) and also after the industrial revolution. These facts must be accepted, and, for this reason, and only for this reason, it is most heuristic to look at cultural evolution in terms of these three phases, which are separated from each other by the neolithic and industrial revolutions.

This approach is merely a conceptual device and is totally heuristic. The descriptions are probabilistic rather than deterministic. They are relative rather than absolute. For example, hunter-gatherer society was simple only in comparison with the later forms of life, and, as archaeologists point out, hunter-gatherer societies could become quite complex, depending on as many as 18 factors.[2]

THE LIFE OF HUNTERS AND GATHERERS

Physical anthropologists generally agree that the modern form of *Homo sapiens* (*Homo sapiens sapiens*) emerged at least 35,000 years ago. Until about 10,000 years ago, when a new form of life developed through the domestication of animals and cultivation of plants, *Homo sapiens sapiens* had been hunters and gatherers.

It is difficult to estimate the conditions of their life exactly, but we may be able to infer this on the basis of the knowledge we have about recent hunters and gatherers. There has been some general agreement about the life of hunters and gatherers. It is commonly assumed that, for them, the largest social unit is the tribe, in which everyone speaks the same dialect of a language, and that the size of a dialect tribe is around 500 individuals, although considerable variation is possible.[3] The dialect tribe of 500 individuals is also the size of the breeding unit.

In terms of population genetics, when the number of tribal members is much less than 500, more women will be brought in from outside the tribe than is customary, and this tends to extend the size of the tribe as a breeding population, making it approximate 500. In contrast, when a dialect tribe includes significantly more than 500 individuals, there will be less need to bring women in from outside the tribal boundaries, and there will be more intratribal mating. When the size of the tribe is very large, physical distance tends to isolate its members, and this contributes to bringing the size of the breeding population closer to 500.[4] Within the dialect tribe, there are local groups, each of which is made up of genealogically related families.[5] In a dialect tribe of 500 individuals, however, not everyone is actively involved in breeding. Within such a population, the size of the effective breeding population has been estimated to be about 175 individuals or less.[6]

In these estimates, the figures are very small, and even if we assume a wide margin of variation, the size of the effective breeding population is so small that physical variation among its members could never have reached the magnitude of racial or ethnic variation. The individuals in a dialect tribe were likely to be very similar to each other in physical appearance. Further, this factor was likely to have been enhanced by polygamy. For example, according to anthropologist James V. Neel and his coworkers, among the Xavante Indians of Brazil, one chief had 5 wives and 23 surviving offspring. Another man had 4 wives and 6 offspring; 4 men had 3 wives each and a total of 13 offspring; 10 men had 2 wives each and a total of 23 offspring; and 21 men had 1 wife each and a total of 24 offspring. Altogether, there were 89 offspring. This means that only one man had produced over 25 percent of the surviving offspring.[7] When such phenomenon exists, the individuals in the tribe naturally become even more similar to each other. It is clear that the smaller the size of the effective breeding population, the more likley this phenomenon is.

In view of the existence of culture among species before us such as *Homo habilis* and *Homo erectus*, there is no doubt that the earliest *Homo sapiens* had well-established culture. Since we know that even chimpanzees show cultural differences among different populations,[8] there is also no question about the existence of cultural variation from one breeding population to another among our earliest ancestors. Yet, as chimpanzee data indicate,[9] at the simplest level of technology variation is a matter of degree, and it is likely that available objects for manipulation were limited by this factor. It is impossible to infer the extent of ideological variation among our earliest ancestors, but in view of the possible role of technology in influencing and stimulating the development of the ideological aspect of culture, at the simple level of technology, their ideology was also likely to have been relatively simple.

It is commonly believed that *Homo sapiens neanderthalensis*, a subspecies of *Homo sapiens* that existed before the emergence of *Homo sapiens sapiens*, had the psychological capacity for sympathy and compassion as inferred from their burial practices.[10] *Homo sapiens sapiens*, then, must have an at least equally developed psychological capacity even from earliest times. Their brain capacity is greater than that of *Homo sapiens neanderthalensis* by about 100 cc (1,300 cc versus 1,400 cc).[11] Indeed, biologically speaking, *Homo sapiens sapiens* 35,000 years ago was the same as us, including the brain, and the potential for psychological variation must have been as great as ours. They were also likely to have shown considerable personality differences.

But one thing we should note here is that psychological variation is often expressed only in response to physical and cultural variations. For example, psychological responses for or against apartheid, anti-Semitism, totalitarianism, or Islamic fundamentalism are possible only when physical or cultural variations exist in reality, when they make people choose positions in one way or another. Without racial, ethnic, and cultural differences in the objective world, it is impossible to develop these attitudes. Thus, despite their biological potential for manifesting deviant or extreme views, hunters and gatherers lived and died without having the opportunity to do so.

Regarding the early forms of *Homo sapiens sapiens*, then, we may assume that they were relatively limited and homogeneous in terms of physical, cultural, and psychological variation. Physically, they were considerably homogeneous, due to the small size of the breeding population; culturally, they were considerably homogeneous, too, because their technology was simple and therefore the tools they could make were primitive and the objects they could manipulate were limited. Psychologically, they were relatively homogeneous because physical and cultural variations were limited and did not make them respond to varied physical and cultural variations in a wide variety of ways, although their brains were potentially able to show varied responses.

THE EVOLUTIONARY MEANING OF THE MANIPULATORY DRIVE

I cannot emphasize too much that it was this condition of life for which the evolution of *Homo sapiens sapiens* was adapted. The evolution of the strong manipulatory drive originally adapted possibly to arboreal life was also adapted on the ground by expressing it in making tools and using them effectively in competition with other species on the ground. The inheritance of this drive helped *Homo sapiens sapiens* significantly by letting it develop stimulus seeking behavior further and further beyond biology; *Homo sapiens sapiens* established a new dimension of life among living species by elaborating this behavior as "culture," and as the dominant means to adapt to the environment.

I should add parenthetically at this point that here I am not talking about the so-called "mighty-hunter hypothesis." This image of *Homo sapiens* as a skillful hunting animal with the help of its culture has been emphasized both by popular writers such as Robert Ardrey[12] and by anthropologists such as Lionel Tiger.[13] But this probably glorifies the capacity of *Homo sapiens* to hunt too much and is therefore unrealistic. The carnivores that we commonly think of as good hunters are not truly hunters but scavengers, most of the time. Besides, they usually attack the old, sick, or very young members of a herd.

This fact is indeed in agreement with the mechanism of evolution because the majority of individuals must be successful in surviving the threat of carnivores, and, in fact, they are. Otherwise they would have become extinct a long time ago, and this in turn means that the carnivores would have become extinct too: for they would have been left without food. This is the true story of carnivores. Compared with them, *Homo sapiens* is much less adapted to such a way of life, and, even with the help of culture, hunting was never as important in human society as some like to believe.

By means of culture, in any event, Homo sapiens could spread from the original habitat, presumably Africa, to other regions of the world, resulting in regional adaptation to the geographical and physical conditions of each area. This created variations among *Homo sapiens sapiens* in the form of racial and ethnic differences. The emergence of this meant even more success for the species in its survival on the face of earth.

Through this process, then, the physical variation of *Homo sapiens sapiens* was created. But this does not necessarily mean that racial and ethnic differences became the source of consistent conflict as they are for us now. The development of racial and ethnic groups as separate breeding populations does not take place without a significant absence of contact between them, often brought about by geographical separation and isolation. The evolution of racial and ethnic groups itself suggests the relative absence of contact between breeding populations.

In this sense, the existence of physical variation in the form of racial and ethnic differences was not a problem for most individuals, because they lived in relatively small, relatively isolated subgroups of the total breeding population, as we can infer from the observation of various primate species. Certainly, it was most unlikely that individuals from different racial groups lived together in a small functioning group such as the family or extended kinship group. Even though the possibility that an individual from another racial group might introduce significantly different genetic material into a breeding population from time to time cannot be completely ruled out, after a few generations his or her phenotype would have disappeared. Since phenotype is the major source of conflict in human interactions, casual interbreeding was not likley to be a problem, unless detailed geneologies were kept by the individuals concerned. This possibility is also very remote.

This is a conceivable reconstruction of life among *Homo sapiens sapiens* before the so-called neolithic revolution, which took place about 10,000 years ago or earlier. Our ancestors no doubt had a strong manipulatory drive, as we do now, but the tools they produced were primitive, and the material objects that could be manipulated were limited. With the primitive level of technology, life was very much the same for everyone, and this meant little or no cultural difference within the group. This condition also suggests that the number of symbolic objects to be manipulated was also very limited. Since the factor of physical variation was not likely to be significant within the breeding population, and since cultural variation was also minimal within it, the potential for psychological variation had a very limited chance to manifest itself.

As a whole, then, any variation among our ancestors before the neolithic revolution was very minor physically, psychologically, and culturally within the unit of their social activity, which would have been the extended family group for most individuals.

THE NEOLITHIC REVOLUTION AND THE
EMERGENCE OF AGRICULTURAL SOCIETY

The neolithic revolution took place in the so-called nuclear areas in western and southeast Asia about 10,000 years ago or earlier, and later, independently, in central America. Although the neolithic revolution refers to a complex of several significant innovations, the two key evolutionary events to change human history were the domestication of animals and the cultivation of plants. From the centers of these innovations, knowledge diffused out over the face of earth to most people.[14]

When the cultivation of plants became established as the predominant way of life in the form of agriculture, an event usually accompanied by the domestication of animals, a different form of life emerged. The village became the unit of life. This is what sociologists and anthropologists consider to be a major way of life in human history, in sharp contrast to contemporary, industrialized, urban, and complex society.

Many names have been coined in order to refer to the traditional, agricultural societies that filled most of our written history. By and large, sociologists and anthropologists agree as to the characteristics of agricultural society, and they use different names to describe the same thing. According to them, agricultural society is tradition-oriented; its people are controlled by informal sanctions such as gossip; social relationships are intimate and personal; there is little division of labor; social structure is rigid with clear class distinctions; and people are ethnocentric and suspicious of outsiders.[15]

The culture of such society may be described as relatively homogeneous, because the village is more or less self-contained and excludes outsiders. In exceptional cases, there may be a racial or ethnic minority within or near the village. But because of rigid social distinctions mainly in the form of class differences, contact with them is relatively limited and is more formal, mainly in connection with trade and business transactions.

Certainly, compared with the situation before the neolithic revolution, cultural variation within society was likely to be greater, and physical variation as well, once there was the possibility for contact with other racial or ethnic groups. This meant, further, that the possibility for psychological variation became greater, compared with people before the neolithic revolution. It is conceivable that the observation of cultural variation as seen in class and occupational differences in the village as well as that of physical variation in the form of racial or ethnic differences may have created a greater range of psychological responses among members of a village.

But there was also a built-in mechanism to counteract this in agricultural society. The strong pressure for conformity by means of informal sanctions based on face-to-face contact made psychological variation very difficult. Also, the rigid structure of agricultural society kept the emergence of the feeling of relative deprivation, for example, to a minimum. When no possibility for achievement or change was visible, people were not likely to feel deprived, even when they saw the system as unjust. Thus, despite the potential for greater variations in physical, psychological, and cultural dimensions, life in agricultural society was relatively homogeneous.

THE IDEOLOGY OF "LIMITED GOOD"

Anthropologist George M. Foster was impressed by the common characteristics found among peasants in various pre-industrial, agricultural societies. He suggested that the mentality of peasants might be expressed as "the image of limited good." According to Foster, this is a common theme in such societies.

In Foster's view, peasants believe that the desired things in life such as land, wealth, health, friendship, love, manliness and honor, respect and status, power and influence, and security and safety exist in finite quantities, and that they are always in short supply. Furthermore, they also believe that a peasant does not have the power to increase the quantities available. The peasant thinks that there are not enough good things to go around because his or her existence is determined and limited by the natural and social resources of the immediate environment.

Foster further argues that when a peasant believes that good things in life are finite and always in short supply within a closed system in his or her narrow environment, it follows that an individual or a family can improve its lot only at the expense of others. An apparent relative improvement for one person is viewed as a threat to the entire community. This perspective is recognizable in the following aspects: (1) the economy, (2) interpersonal relationships, (3) health, and (4) masculine honor.

The economy of peasant life is not productive, because land is usually limited, and, furthermore, land becomes more and more limited as the population expands and the soil deteriorates. In interpersonal relationships, a peasant assumes that friendship, love, and affection are limited. As a result, a peasant must avoid showing excessive favor or friendship. Sibling rivalry is caused because even maternal love is limited. A husband is jealous of his son and angry with his wife for the same reason.

Health, too, is limited in quantity. Blood is nonregenerative. Blood may be equated with semen, and the exercise of masculine vitality is seen as a permanently debilitating act. Sexual moderation and the avoidance of bloodletting are important. Even a woman's long hair may become a source of apprehension because she may lose her vigor and strength by having long hair.

Honor and manliness, too, exist in limited quantities. Real or imagined insults to personal honor must be vigorously counterattacked because honor is limited, and a peasant cannot afford to lose it.

When good things in the environment are assumed to be limited, and when personal gain can only occur at the expense of others, the maintenance of the status quo is the most sensible way to live, because to make economic progress or to acquire a disproportionate amount of good things is a threat to the stability of the community. Stability is maintained by an agreed-upon, socially acceptable, preferred norm of behavior, and sanctions and rewards are used to ensure that real behavior approximates the norm.

As a result, there is a strong desire to look and act like everyone else and to be inconspicuous in position and behavior. For the same reason, a peasant is reluctant to accept leadership roles. The ideal peasant strives for moderation and equality in his or her behavior. If a peasant should behave excessively, then gossip, slander, backbiting, character assassination, witchcraft or the threat of it, and even actual physical aggression is used by the rest of society against such a person.[16]

It is difficult to say to what extent this generalization applies to people after the neolithic revolution and before the industrial revolution. Foster's work is based on various studies of peasant societies in the contemporary world at different stages of development, and if the picture Foster presents approximates the condition of people between the neolithic and industrial revolutions, we may infer that despite their biological potential, their psychological variation was limited. In many agricultural societies, physical and cultural variations were likely to be considerably greater than in hunter-gatherer societies. Yet if people were obsessed with the belief of "limited good" and thought and behaved like everyone else, their psychological variations might not have been much greater than those among hunter-gatherers.

THE INDUSTRIAL REVOLUTION AND THE DOMINANCE OF TECHNOLOGY

The ultimate source in changing this way of life was the development of technology. This was brought about by what is called the industrial revolution. The drive to manipulate more and more of the environment made people improve their tools and techniques for using them, if we use the Aristotelian scheme of technology. As a result of the industrial revolution, a totally new form of society called industrial society emerged, in which most of the characteristics associated with agricultural society were reversed. And in industrial society, the potential sources for physical, psychological, and cultural variations that had been dormant in agricultural society were activated.

As in the case of agricultural society, there is general agreement among sociologists and anthropologists as to the common characteristics of industrial society. According to them, industrial society is characterized by such features as a future-oriented mentality, always assuming the possibility for change; social control by the formal means of laws applicable equally to all; more casual and temporary interhuman relations; an open class structure with the possibility for mobility from one class to another; and more openness to foreign ideas and to outsiders.

Characteristics of agricultural society, such as strong pressure for conformity and a rigid social structure, which prevented the possible development of, for example, the emotion of relative deprivation, were replaced in industrial society by contrary characteristics. As a result, human variation in physical, psychological, and cultural dimensions became apparent. The possibility for more contact between different racial and ethnic groups increased. Women began to invade men's occupations. Contacts between different cultures increased, and, within the same society, contact between subcultures became a fact that could not be avoided. Different views began to be expressed openly. The emergence of industrial society meant the realization of both extremes of the bell-shaped curve of human variation in the physical, psychological, and cultural dimensions.

Of course, this took place gradually through time, and, even in the most open and tolerant societies such as Sweden, not all forms of extreme ideas and behaviors are legal. For example, terrorism and the sexual exploitation of children are illegal. Yet, compared with agricultural society, industrial society displays far more extensive human variation.

Through time, technology in the sense of the tool as well as in that of the technique for using it evolved much more extensively and powerfully than could be imagined at the beginning of the industrial revolution. As the social philosophers of the Frankfurt school such as Herbert Marcuse point out, technology has grown so powerful through time that it now dominates both nature and us.[17]

This can be understood, at least in part, by looking at our biological background in two ways. First, there is no question about the role of the cerebral cortex in the human brain: it has made us very intelligent and has permitted us to make more and more efficient and powerful tools and to use them effectively. Second, reflecting the evolutionary past of the primates, there is a very obvious tendency for people to conceive of their manipulatory behavior as a matter with definite bounds and not visualize possible consequences beyond those bounds. If I may borrow G. H. Mead's expression, this deals with the "manipulatory area." As you will recall, according to Mead, this is a physical space that is subjectively established by the individual through recognizing the results of his or her attempts to manipulate the environment. When such attempts turn out to be successful, in the sense that the individual can deal with the environment in the way he or she intended, the object that has been manipulated remains within the manipulatory area of that particular individual.[18]

TECHNOLOGY FOR MANIPULATION

As primates, we deal with the environment from the perspective of the individual and this way of dealing with the world is genetically acquired. But, as a result of the development of the capacity to symbolize, a person can also identify with another person and vicariously experience the manipulatory behavior of that person. Of course, people do not identify with just any person. Most typically, one identifies with someone who can successfully manipulate the object that one wishes to manipulate but cannot. Examples include the son's identification with the father in the Oedipus conflict, or admiration of a professional athlete, singer, conductor, politician, or commander-in-chief, who are skillful at expressing the manipulatory drive successfully. Technology in human culture has evolved essentially on these two bases: (1) the individual's perspective and (2) identification.

If we understand technology's meaning in human history, it is easy to see why it has been the dominant force in shaping, directing, and even determining the course of history and civilization. Weapons are used against animals to be hunted, and against enemies to be defeated or exterminated. There is no reason to identify with an animal to eat or an individual of the enemy group; they are objects to be manipulated. At the same time, it is quite logical and natural to identify with a successful hunter, warrior, or political leader through whom one can vicariously manipulate objects in the manipulatory area.

To be attacked, captured, tortured, or killed by the enemy is not a pleasant experience, but it is important to remember that a person in this situation is merely an object to be manipulated in the manipulatory area. In this sense, a war is a very rewarding experience—if you win. It is also easy to understand why a nation, when it suffers from conflict and divided opinions among its people, can be unified merely by starting a war: everyone can have the opportunity to express his or her manipulatory drive as much as he or she wishes to, or to experience it vicariously by identifying with the nation's soldiers, heroes, and politicians.

Although war is an extreme example, similar consequences of technology have been abundantly seen in human history. In accordance with the development of technology through time, this phenomenon is more and more apparent, and is becoming extensive. Take the problem of environmental pollution and destruction, for example. People drive automobiles, and many feel that to drive a car is a basic human right, something of supreme importance. They violently resist any attempt to limit parking or to lower the speed limit in order to reduce air pollution. These restrictions run directly counter to our drive to manipulate. By driving a car, one can express one's manipulatory drive in two ways: both the car and one's own body are manipulated.

Indeed, in the contemporary industrial world, where the majority of people are mere cogs in bureaucratic systems and have very limited opportunities to express their manipulatory drives, there are not many things that one can manipulate freely at will. As C. Wright Mills points out, in mass society even communication is one-directional, and an individual in "democracy" cannot make his or her opinion heard.[19] If one is lucky, one may be able to manipulate symbols such as power, money, or prestige, but these means are closed for most individuals.

What most individuals can do instead, in addition to vicarious manipulation through identification is to try to manipulate other human beings by abusing women, children, subordinates, or members of a minority group; or the material self, through jogging, yoga, or the use of drugs; or material objects of technological invention, such as computers, cars, or stereo systems. For example, one drives a car at speeds far beyond the speed limit not in order to reach one's destination earlier, but merely to derive a sensation of mastery from the experience.

For young people, to listen to excessively noisy rock music and its destructive rhythm is a vicarious way of satisfying their manipulatory drive. As a matter of fact, there are rock musicians who perform violent or destructive acts on stage, and this clearly shows the need for the vicarious satisfaction of the manipulatory drive among the audience. Violent video films have the same function.

For insecure individuals in mass society who are not satisfied with today's society but do not dare to be deviant by expressing the manipulatory drive as they please, a mass rock concert or a mass marathon quite neatly fulfills the needs of being in a crowd and of expressing the manipulatory drive at the same time. In these situations, one accepts the miserable fact of being a mere item with a number in the contemporary industrial society and at the same time tries to satisfy one's manipulatory drive in an approved manner, either directly or vicariously.

Even such deviant phenomena as arson and anorexia nervosa may possibly involve an unusual expression of the manipulatory drive for a powerless and insignificant being in industrial society: it is conceivable that one derives the satisfaction of mastery by setting a fire or by seeing the transformation of one's own body.

Indeed, there is not much we can do in the contemporary world of industrial society when we want to express our manipulatory drive. The behaviors that I have listed here are conceivable alternatives for the masses in most cases. When contemporary industrial society is characterized by capitalism, technology, a consumption-oriented economy, and bureau-cracy, most individuals have very little opportunity to express their manipulatory drives.

People know well that air pollution is harmful to practically all living species, including ourselves. Yet the harmful effect upon one's own health is not immediately visible. Insofar as one can feel satisfied by seeing the successful manipulation of the car and one's own body without any immediately negative effect, there is no reason to stop driving the car. The reward of driving the car is too precious for the powerless and miserable masses.

The same observation can be made for many smokers. Here, the smoker manipulates nicotine as the tool, and his or her own body is the material to work on. The end result is an ecstatic experience involving the action of neurotransmitters. Through this intended result, one feels one has successfully manipulated one's own body. In the Aristotelian scheme, the material to work on is at the same time the tool-user. Except possibly for those smokers who wish to stop smoking for good and envy nonsmokers and yet cannot stop smoking, smokers have no reason to identify with nonsmokers, because, evidently, nonsmokers do not manipulate any object successfully that smokers wish to manipulate. The condition of health can be seen as a desirable object to be manipulated even for smokers, but as long as they do not experience any serious illness caused by smoking, the reward of manipulating their own bodies is preferable to it.

In these examples, it is clear that one sees the environment in terms of satisfying one's own manipulatory drive directly or indirectly. Insofar as one can do so without being disturbed, one can and will continue the same way of satisfying the drive. This is a direct reflection of our primatological background, in the sense that one can see the consequences only within the manipulatory area. At the same time, this is the very source of problems in industrial society.

OUR PRIMATOLOGICAL BACKGROUND AND THE PROBLEMS OF INDUSTRIAL SOCIETY

The manipulatory drive, then, has been the biological basis for the development of technology and industrialization in human history. As a species belonging to the order called Primate, we invented and improved tools. The tools were applied to the environment, and the result was seen as satisfying and rewarding, and so this activity continued. Of course, most individuals in human history never invented tools, but the idea of new technology was extensively borrowed and copied throughout human history, as anthropologists repeatedly tell us.[20]

In the contemporary industrial world, technological innovation is created basically within the bureaucratic system. In the capitalist West, the rewarding object to be reached for and manipulated is profit, and the inventor, engineer, or researcher sees honor and prestige as well as monetary reward as desirable objects. The president of the company or the head of the laboratory in turn sees such a person as the tool through which he or she can reach for profit for the company or the laboratory. In the East European communist system before 1989, the monetary aspect may have been less important, but the arrangement for the manipulatory attempt was very much the same. A person engages in a research project in which the environment is to be manipulated in a new way. When successful, one makes an invention, innovation, or discovery, and, along with it, symbolic objects such as honor, prestige, or money come along.

In this sense, the pattern of technological development has been very much the same regardless of the political or economic system of the industrial country. The awareness of the consequence of a new technological innovation has been limited to the apparent effect of reaching for and manipulating the desired object, without paying attention to consequence beyond the phenomena within the manipulatory area. But pollution and environmental destruction are happening precisely because of this. This has been the history of industrialization most of the time. As a matter of fact, such words as "ecology," "pollution," or "environment" began to be used commonly only in the 1960s in the West and in the 1980s in Eastern Europe.

Certainly, we may try to know all the consequences and implications of an innovation. Nowadays, a business firm usually prepares a feasibility study before it decides whether or not to market a new product. But the aim of a feasibility study is basically a study of profit potential, and other possibilities are clearly secondary. Similar studies are also made by the public sector, in which more emphasis is placed on all possible consequences in examining a technological innovation. But, in reality, it is impossible to know all the consequences of an innovation.

Sociologist Robert K. Merton tells us that an action with a clearly defined purpose may bring about totally unintended results. First, an unintended result occurs when the set of consequences of any repeated act is not constant but instead there is a range of possible sets of consequences, any one of which may follow the act in any given case. If the person does not know about this range, but simply knows of the existence of only one set of consequences, unexpected results might occur. Second, there is a factor of error. An error may occur in (1) the appraisal of the present situation on which the person's decision is based; (2) the inference from the appraisal of the present situation to the future situation; (3) the selection of a course of action; or (4) the execution of the chosen set.

Third, the person's paramount concern with the foreseen immediate consequences excludes his or her consideration of further or other consequences of the same act. Fourth, basic values compel the person to act in a certain way without considering the further consequences of the act. Fifth, public predictions of the future can change the outcome because the prediction becomes an element in shaping the direction of the action.[21]

These unintended and unexpected consequences of an act are quite natural if we remember our biological background as a primate species. The belief that the human being is a rational animal is false and is a product of categorical thinking. A person may be rational at one particular time in response to one specific situation, but a person can never be rational enough to be able to recognize all the consequences of an act. Science and technology developed mostly on the basis of this false assumption, which was derived from the philosophy of the Enlightenment.

The history of biological and cultural evolution in *Homo sapiens sapiens* is a complex matter. But, at the same time, surprisingly enough, it can be summarized in terms of four distinct characteristics. First, there are two biological phenomena that are given to us, so to speak, to begin with. These two phenomena have been the main topics of discussion in this book thus far, namely (1) stimulus seeking behavior and the corresponding manipulatory drive, and (2) variation.

Second, these two forms of biological phenomena are significant advantages for *Homo sapiens sapiens* in terms of biological evolution. But, as a result of cultural evolution, which was brought about by biological evolution, these advantages are, ironically, hurting us to such an extent that we may become extinct. This applies to both forms of phenomena. Namely, the drive to manipulate is so strong that we create more and more tools for manipulation that become more and more efficient and powerful. This can be seen in the development of weapons as well as the destruction of the global environment.

Similarly, physical, psychological, and cultural variations were originally adaptive because they were to function as "banks" to supply materials to deal with the physical environment by maintaining varied genes, ideas, and approaches. When the physical environment changes, variations are there to help us to adapt better to changed conditions. But instead of understanding and appreciating variation, we create conflict by rejecting it. Physical, psychological, and cultural variations in the world have been constant sources for major and minor conflicts of all forms.

Third, as a result, both forms of biological phenomena are not at all helping us. To make it worse, we suffer from these phenomena. We become frustrated, dissatisfied, alienated, and irritated. We become angry and sad.

Fourth, mainly due to the development of more and more efficient tools to deal with the environment, combined with more varied opinions and actions in using these tools, the problem is becoming more and more serious, including the possibility for our self-extermination as a biological species.

NOTES

1. Leslie A. White, *The Science of Culture* (New York: Grove Press, 1949), p. 368.

2. T. Douglas Price and James A. Brown, "Aspects of Hunter-Gatherer Complexity," in *Preshistoric Hunter-Gatherers: The Emergence of Cultural Complexity*, ed. T. Douglas Price and James A. Brown (Orlando, Fla.: Academic Press, 1985), p. 7.

3. Joseph B. Birdsell, "Some Predictions for the Pleistocene Based on Equilibrium Systems among Recent Hunter-Gatherers," in *Man the Hunter*, ed. Richard B. Lee and Irven DeVore (Chicago: Aldine, 1968), pp. 230, 232, 233, 237, 239; Joseph B. Birdsell, The Magic Numbers '25' and '500': Determinants of Group Size in Modern and Pleistocene Hunters," in Lee and DeVore, eds., *Man the Hunter*, p. 246; Sherwood L. Washburn and C. S. Lancaster, "The Evolution of Hunting," in Lee and DeVore, eds., *Man the Hunter*, p. 302.

4. Birdsell, "Some Predictions for the Pleistocene Based on Equilibrium Systems among Recent Hunter-Gatherers," pp. 237–38.

5. Ibid., pp. 234, 235, 239.

6. Ibid., pp. 238, 239.

7. James V. Neel, F. M. Salzano, P. C. Junqueira, F. Keiter, and D. Maybury-Lewis, "Studies on the Xavante Indians of the Brazilian Mato Grosso," *American Journal of Human Genetics*, 16 (1964), pp. 52–140.

8. W. C. McGrew, C.E.G. Tutin, and P. J. Baldwin, "Chimpanzees, Tools, and Termites: Cross-Cultural Comparisons of Senegal, Tanzania, and Rio Muni," *Man*, 14 (1979), pp. 185–214.

9. Ibid., pp. 185–214.

10. F. Clark Howell, *Early Man* (New York: Time-Life Books, 1976), pp. 128–30.

11. Kathleen Gibson, "Has the Evolution of Intelligence Stagnated since Neanderthal Man?" in *Evolution and Developmental Psychology*, ed. Julie Rutkowska and Michael Scaife (Brighton: Harvester Press, 1985), p. 104.

12. Robert Ardrey, *African Genesis* (New York: Dell, 1966); Robert Ardrey, *Territorial Imperative* (New York: Dell, 1968).

13. Lionel Tiger, *Men in Groups* (New York: Random House, 1969).

14. C. C. Lamberg-Karlovsky, ed., *Hunters, Farmers, and Civilizations* (San Francisco: Freeman, 1979).

15. Harold M. Hodges, Jr., *Conflict and Consensus: An Introduction to Sociology* (New York: Harper and Row, 1971), pp. 148–50.

16. George M. Foster, "Peasant Society and the Image of Limited Good," *American Anthropologist*, 67 (1965), pp. 293–315.

17. Herbert Marcuse, *One Dimensional Man* (London: Routledge and Kegan Paul, 1964), Chapter 6.

18. George Herbert Mead, *The Philosophy of the Act* (Chicago: University of Chicago Press, 1938), p. 141.

19. C. Wright Mills, *The Power Elite* (New York: Oxford University Press, 1959), p. 304.

20. Alfred L. Kroeber, *Anthropology*, new ed. (New York: Harcourt, Brace and Co., 1948).

21. Robert K. Merton, "The Unanticipated Consequences of Purposive Social Action," *American Sociological Review*, 1 (1936), pp. 894–921.

8

THE BIRTH AND DEATH OF
LIBERALISM

After the disappearance of *Homo sapiens neanderthalensis* about 35,000 to 40,000 years ago, only one variety of *Homo sapiens*, namely *Homo sapiens sapiens*, continued to exist, and that is us. Biologically, *Homo sapiens sapiens*, including its brain structure, has remained the same for at least 35,000 years. This means that our ancestors who lived by hunting and gathering until the neolithic revolution could think and feel just like us, theoretically speaking. They had the same potential for psychological variation as we have now in industrial society.

Yet because of their limited physical and cultural variation, their potential for psychological variation was not activated, due to our tendency to think and feel in response to what we see and experience. When the hunters and gatherers were physically similar to each other and were surrounded by their simple culture, there was little possibility for thinking and feeling in a very varied manner.

After the neolithic revolution, when our ancestors began to settle and to live a new life based on domesticated animals and cultivated plants, their physical and cultural variations became greater. This change meant a greater possibility for more psychological variation among them. Yet a large number of studies by sociologists and anthropologists about preindustrial societies and peasant cultures consistently indicate a strong tendency for constraining psychological variation among people. As we have seen in Chapter 7, one possible explanation has been summarized by anthropologist George M. Foster by the expression "image of limited good."

THE BIRTH OF LIBERALISM

After the industrial revolution, practically all of the characteristics associated with agricultural society were reversed. No doubt the most spectacular example, philosophically speaking, is the emergence of a totally new perspective about ourselves, in which a more objective observation of human beings began to appear. For example, Thomas Hobbes and Jeremy Bentham emphasized the importance of pleasure and pain in our behavior, while A.R.J. Turgot and Marquis de Condorcet believed in our ability for progress.

But above all, by emphasizing such ideas as equality, humanity, reason, and compassion, the philosophy of the Enlightenment encouraged the acceptance of human variation; two of the consequences of this way of thinking were opposition to slavery and support of more humane treatment for the mentally ill. Of course, fragmentary ideas of this nature existed earlier in human history. For example, Socrates valued truth at the cost of his life, and the Stoics emphasized the importance of individuality in the tradition of Western philosophy. In the East, Confucius taught the value of humanity to his disciples. However, only in the modern period of Western thought did these fragmentary ideas become more or less unified to emerge as the philosophy of the Enlightenment.

The ideology of liberalism and individualism is one of the important products of the philosophy of the Enlightenment, in which the value of the individual and variation among us are accepted as conditions of being human. In the history of Western philosophy, such philosophers as Descartes, Milton, and Spinoza were instrumental in creating the condition for this perspective. They were followed by many Enlightenment philosophers, such as Voltaire, Locke, Rousseau, Vico, and Condorcet.

In this way, liberalism as a system of thought emerged for the first time in human history. The acceptance of variation within a single people and also of peoples differing from each other in racial, ethnic, linguistic, or religious respects was embodied in action, as in the development of interest in "exotic" peoples and cultures in arts and science, the independence of the Americas, the liberation of the Jews, the abolition of slavery, and suffrage for women. In this sense, the ideology of liberalism is in accordance with the phenomenon of our biological background in that it accepts and recognizes physical, psychological, and cultural variations. This is indeed the most realistic philosophy in human history for building culture and civilization. Or, to put it more accurately, this could have been the most realistic philosophy if the philosophy of the Enlightenment had been a completely consistent ideological system, as I shall explain shortly.

LIBERALISM AND CAPITALISM

Capitalism may be considered to be another product of the philosophy of the Enlightenment, in the sense that it is a way of dealing with the environment by emphasizing rationality, progress, free enterprise, reason, and empiricism. When these ideas are focused on a specialized form of activity that involves resources in its search for profit, it becomes capitalism.

In the beginning, there was no logical inconsistency between capitalism and liberalism. After all, both were products of the same philosophy, and both were in principle based on the same set of assumptions. But gradually capitalism began to destroy liberalism. This has been apparent in two ways.

First, in order to engage in investment, production, distribution, advertisement, and sales, or, in brief, in order to carry out capitalist activities, capitalists came to rely more and more on technology. This is quite logical. Since the capitalist principle prescribes that profit will be attained by minimizing input and maximizing output in the form of profit, it is necessary to have as efficient a tool as possible to do this. This can be done by having better technology. Thus, the development of technology and the growth of capitalism have taken place in close coordination; capitalism required better technology, and better technology brought about more profit, which in turn encouraged more improvement in technology.

From the standpoint of liberalism, technology *per se* is neither good nor bad. Indeed, there are several characteristics of the philosophy of the Enlightenment that both liberalism and capitalism share, and, insofar as these characteristics are concerned, there is no logical inconsistency. Yet when technology is applied to the environment as a tool, the tool-user is a human being, and one human being may significantly differ from another human being. The potential source of incompatibility lies here. That is, the way technology is used may not necessarily be acceptable to other human beings. When one individual uses technology in a certain way, the result may be disturbing to another individual, and from the liberal point of view this is objectionable and unacceptable.

Second, capitalism began indirectly to destroy liberalism via communism and socialism. As we all know, capitalism was instrumental in the rise of communism and socialism, and these ideologies are incompatible with liberalism because of their emphasis on the state as the ultimate authority over the individual. In these ideologies, the state has the right to control freedom of speech and the press, ownership, and, above all, the means of production, and these characteristics are incompatible with liberalism. Of course, communism and socialism, too, emphasize the importance of technology as a means of manipulating the environment in general and as a means of production in particular, and the problem of technology against liberalism is also present in societies based on communism or socialism. The state can exercise technology freely by prohibiting opposition, and this is carried out under the assumption that what the state does is good for the people—a view that totally ignores variations of individuals within society.

Thus, ironically, liberalism has been mistreated by capitalism, its ideological sibling, directly and also indirectly via communism and socialism. Both liberalism and capitalism are products of the philosophy of the Enlightenment, but the emphasis on and development of technology resulted in the destruction of liberalism. When capitalism is combined with the political system of democracy as in the case of Western Europe and North America, liberalism is additionally disintegrated by the tyranny of the masses, which in so many situations simply ignores the voices of the minorities, as pointed out by critics of democracy such as Tocqueville.

In this sense, contrary to what some like to believe, it is not just communism and socialism that have been a threat to liberalism. Capitalism, too, is equally fatal for it. After all, whichever political system one is in, if the power of technology is in the hands of someone who does not think like oneself, there is always a threat to liberalism. The ownership of the means of production does not make any difference in this regard.

An ideological synthesis of capitalism and socialism, as in the case of social democracy in Sweden, is not favorable to liberalism either. By following the ideology of socialism, Swedish social democracy makes an idol of underdog mentality, almost always supporting the stigmatized side in society. For example, crimes against a person entail little or no punishment, and it is often the victims who must suffer without support or compensation. In effect, this is as bad as totalitarian dictatorship, because, in effect, it is the offenders, the extreme individuals, who dominate the rest of the population.

NEGATIVE FREEDOM

Philosopher Isaiah Berlin's well-known work on freedom symbolically represents the defeat of classical liberalism. In a paper entitled "Two Concepts of Liberty," Berlin distinguishes between "negative" and "positive" liberty. Berlin thinks freedom and liberty are synonymous, and here I shall do so as well.

In Berlin's opinion, "negative" freedom is the kind of freedom emphasized and discussed by classical English political philosophers such as John Locke and J. S. Mill. This is the orthodox form of liberalism, in which freedom means the freedom not to suffer from the interference of others. According to this form of freedom, one should be entitled to a certain minimum amount of personal freedom, which must not be violated.[1]

In contrast, "positive" freedom is freedom to be a subject and doer. This means the freedom to be one's own master. One has this freedom when one conceives goals and policies of one's own and is able to realize them.[2]

Berlin himself likes negative freedom. He states: "No doubt every interpretation of the word liberty . . . must include a minimum of what I have called 'negative' liberty."[3] But, at the same time, he is pessimistic about the practicality of negative freedom. He thinks that the two forms of freedom historically developed in divergent directions, and not always by logically reputable steps, until they came into direct conflict with each other.[4] For this reason, Berlin tries to solve the problem by making the issue relativistic. That is, Berlin points out that the idea of negative freedom is not something deeply rooted in the history of Western civilization.[5] For Berlin, negative freedom is not absolute but relative; it depends on such factors as culture, time,[6] and level of education.[7]

Berlin also rejects retreatism, such as the self-emancipation of ascetics, stoics, or Buddhist sages, because he thinks this is just the result of "sour grapes."[8] Berlin's suggestion is to accept that human goals are many, that not all of them are commensurable, and that many are in perpetual rivalry with one another.[9]

In essence, what Berlin does in his paper is to criticize the idea of negative freedom, especially the version advocated by John Stuart Mill. But Berlin himself does not present any satisfactory alternative. It is certainly true that the idea of negative freedom is not found universally in all cultures nor among all individuals. It is probably correct to say that "the domination of this ideal has been the exception rather than the rule, even in the recent history of the West."[10] But this fact itself is not a justifiable reason for abandoning it.

The idea of negative freedom is certainly not what everyone wants everywhere. But neither is it an intellectual product of a few eccentric philosophers in some ivory tower. Rather, it is a by-product of industrialization. As a result of industrialization, the physical and cultural variations in our environment multiplied, and, in response to these variations, our psychological variations also multiplied. Psychological variations in turn stimulated further psychological variations. This is how the idea of negative freedom emerged. This is what philosophers of Romanticism such as

Hegel would call the "unfolding of the mind." More and more ideas emerged in response to the development and subsequent heterogeneity of our physical, psychological, and cultural environments.

Here is the tragic fate of the idea of negative freedom. It emerged in response to the physical and cultural variations in the environment of the West after the industrial revolution. The philosophy of the Enlightenment, which was more or less concurrent with it, paved the way for new ideas, such as the idea of negative freedom. Yet, ironically enough, it was crushed precisely because of physical, psychological, and cultural variations. In the contemporary world of heterogeneity, it is exactly the idea of negative freedom that is most needed. But instead of promoting it even more vigorously than earlier liberals, Berlin abandons it.

NEGATIVE FREEDOM AND TECHNOLOGY

The idea of negative freedom was predominantly academic and a material for intellectual exercise when J. S. Mill and Benjamin Constant wrote about it. But in the world today this problem is more than academic. It is a problem for everyone, not because the ordinary citizen in the street knows what these liberals in the nineteenth century wrote and said, but because everyone in the world is threatened by problems in their environment. Or, to put it more accurately, the individual in the world today is disturbed by the lack of negative freedom to such an extent that he or she cannot maintain his or her "home range" in peace.

The intrusive and disturbing factors include various forms of pollution, such as the pollution of air, land, and water, resulting in the pollution of food and subsequently of the body; noise pollution; the invasion of private information via the computer; commercial pollution; the mass extermination of civil populations in war or by accident; and the imposition of certain political or religious ideas upon everyone in society by force. These problems theoretically apply to anyone in any culture, and, in this sense, the need for negative freedom has become even more urgent now than it was in the time of J. S. Mill.

When such words as "liberty" and "freedom" were used by intellectuals in the eighteenth and nineteenth centuries, they were usually thinking of politics. They wanted to know to what extent the government could limit the freedom of each citizen to exercise his or her rights. But since then, the nature of disturbance and interference of personal freedom has changed significantly. Instead of—or, more correctly, in addition to—political freedom, there has emerged a powerful threat to personal freedom. This is the threat of technology.

If we remember that our stimulus seeking behavior is accompanied by the drive to manipulate, it is easy to understand that technology, as the tool for manipulating our environment, naturally comes into the picture when we discuss personal freedom. There is a limit to deprivations of negative freedom when there is no tool. But by improving the efficiency of the tool, the ways of invading and crushing a person's negative freedom become more and more effective. Unfortunately, this is exactly what has happened and is happening.

In this sense, technology is contributing to the death of liberalism, and, above all, to the death of negative freedom. Technology almost always works to promote "freedom for" at the expense of "freedom from." As a human being of limited knowledge, I am unable to recognize all the consequences of technological development and how it has worked against the individual, especially in crushing his or her negative freedom. But I would like at least to attempt to identify some of them, which in my opinion have particularly grave consequences in our lives.

There are at least six such consequences: (1) technology as a source of conflict, (2) the allocation of more power to single individuals, (3) more power to everyone, (4) more diffusion between cultures and nations, (5) the greater difficulty of avoidance, and (6) the emergence of new subcultures. I shall explain these consequences one by one.

TECHNOLOGY AS A SOURCE OF CONFLICT

A new technological innovation offers a new possibility to manipulate objects that we previously have been unable to manipulate, or to manipulate old manipulatable objects in a different way. Whenever a new technological innovation is feasible or has already appeared as a reality, there almost always emerge opinions for and against it. Many of the problems we have now regionally, nationally, internationally, and globally are of this nature.

Genetic engineering is a good example of a new technological innovation that has become a source of debate by opening up the possibility of manipulating DNA. An example of a debate over a new way of manipulating the old manipulatable objects is seen in an attempt to prohibit advanced nuclear weapons while allowing the conventional ones. The purpose of all weapons is ultimately to kill the members of the enemy nation, and, in this sense, there is no difference between conventional weapons and advanced nuclear weapons. In both situations, a person who happens to become a target is killed, and it makes no difference if the number of people killed is two or two million. But most politicians do not appear to reason in such a way.

When we had only conventional weapons, there were just two categories of people who had intense feelings for or against them. But by
adding the issue of the advanced nuclear weapons, there are three known
categories of people: those who are against both conventional and advanced nuclear weapons; those who are in favor of both conventional and
advanced nuclear weapons; and those who accept conventional weapons
but reject advanced nuclear weapons. Theoretically, there can be the fourth
category of people who are for advanced nuclear weapons but against
conventional weapons, but this group does not seem to exist. The point I
would like to make here is that a new technological innovation can always
be a factor in creating different opinions. Without the innovation, people
do not realize that they would hold different opinions about it.

To have more technological innovation means to increase the possibility
of dividing people into almost unlimited combinations of opinions. By
having advanced nuclear weapons, a third category of people emerged,
who are for conventional weapons but against advanced nuclear weapons.
To have yet another form of weapon means that still more categories of
individuals will be created.

To divide people regarding a new technological innovation results in a
struggle for the power over the manipulation of objects. When one directly
or vicariously succeeds in manipulating the object after a debate, negotia-
tion, or more explicit exercise of power, one's manipulatory drive is
satisfied. But, at the same time, this means that another category of people
who have failed in manipulation exists. To lose in a struggle is a source of
suffering; this happens because a technological innovation forces people
to take a side and to respond to the issue in different ways.

Since people are psychologically varied, and since they are likely to
respond differently to technological innovations, to have more and more
technological innovations in a wide variety of areas means that everyone
in society has a greater and greater possibility of becoming alienated. One
may be totally satisfied with a new technological innovation, but this does
not necessarily mean that the next innovation will also be satisfactory. It
is a matter of time until one comes across an innovation that is manipulated
against one's will by a category of people who derive satisfaction by
exercising their manipulatory drive in the way they want.

In capitalist society, this problem is made worse by the tendency to
create a need for technological innovation among the masses even when
they do not feel a need for it to begin with. Since industry in complex
capitalist society requires huge capital investment and long-range planning,
a company cannot afford to fail in the marketing of a new product. In
"affluent" industrial society, the masses are persuaded to buy the new

product through what economist Galbraith calls "demand management." The purpose of demand management is to ensure, by manipulating attitudes and tastes, that people will buy what is produced.[11]

In this way, many new technological products are put into circulation in society, creating new sources for divided opinions. A few familiar examples are plastic bottles, aluminum cans, mercury batteries, and leaded gasoline, all of which were not "demanded" by consumers in the sense defined by traditional economics. Yet consumers were made to feel that these products were better and more up-to-date. They began to buy. From the standpoint of our environment, these products are all undesirable, yet they have not disappeared, and people are divided for and against them in industrial society.

THE ALLOCATION OF MORE POWER TO SINGLE INDIVIDUALS

In view of the strong manipulatory drive we have, it is understandable that technology is becoming more effective and more powerful in manipulating the environment. What appears to be incidental, yet actually is very important, is that single individuals are also acquiring more and more power to manipulate, along with the development of technology in society as a whole. For example, the method of killing people has clearly become unbelievably effective when we see the evolution of weaponry from the bow and arrow, through the gun, the machine gun, and the bomb, to the nuclear weapon, the laser, and the neutron bomb. In principle, a single individual can theoretically release an advanced weapon as easily as an arrow, and this means that more and more individuals can be affected by a chance factor associated with a single individual. This problem applies not only to weapons, but also to many other technological innovations that can kill people on a large scale. One person's mistake, misunderstanding, or psychological fluctuations in state of mind can affect, damage, or kill a large number of people. The Chernobyl accident of 1986 is perhaps one of the best examples of this.

As we all know now, the accident in Chernobyl resulted from the combination of a reactor design with inherent control problems *and* the reckless and deliberate disregard of established safety procedures by the plant operators. The accident began with the initiation of a test to see how long the steam turbines would run while coasting to a stop when the reactor was suddenly shut off. During the test, almost all the safety features of the reactor were disconnected or turned off. And when the operation encountered serious problems, operators blocked the signals to the emergency reactor protection system. There developed a sharp temperature rise in the fuel channels of the reactor, which made the whole system go out of control. This resulted in two explosions that blew the top off the reactor vessel.[12]

Various accidents involving tankers, computers, or airplanes are not unfamiliar to us, and usually these accidents are caused by only one or a few individuals. Furthermore, it is impossible to have a technological innovation that is 100 percent safe, and, in this sense, we are always surrounded by the possibility of a catastrophe, no matter how slight it might be. This is taken for granted by engineers, but the layman tends to have a greater amount of blind faith in modern science and technology than he or she should. Witness, for example, the way a bank responds when its cash dispenser does not work properly. A common answer the customer gets from the bank is: "The computer does not make a mistake."

There are two factors that we must consider here. First, there is the problem associated with the peculiarity of a single individual in handling and manipulating a complex technological system. An accident can happen when a person fails to manipulate the system properly. Second, another possibility for an accident lies in mistakes an individual may make in responding to an accident or breakdown in a technological system. By not responding to it properly, an individual can create a catastrophe out of a minor accident that could have been prevented by proper and routine measures. Whichever is the case, a single individual can have an incredible amount of power over other human beings, including over their lives.

These cases deal with individuals who are not causing accidents intentionally. But, in view of psychological variation among us, it is also conceivable that a single individual can cause an accident on purpose. Indeed, there are known cases. In 1982, a schizophrenic pilot of Japan Air Lines intentionally crashed his passenger plane, carrying 174 passengers and crew, killing 24 and injuring 150 persons.[13] Indiscriminate killing by a sniper is also a phenomenon that crops up once in a while in various industrial societies.

To explain these incidents away by saying that such people are psychopaths does not solve the problem. Rather, these incidents suggest that we must be prepared to accept that people are psychologically varied to such an extent that a grave accident can happen at any time in a technologically complex society. Terrorists, too, can take advantage of weaknesses in technological society and interrupt the communication network, or explode a bomb in a car, train, or airplane. These familiar incidents result from the combination of psychological variation and the advancement of technology. This means that, if an extremist wants, by relying on advanced technology he or she can carry out a mass murder or even genocide relatively easily.

It is totally unnecessary to go into the details of the Nazi Holocaust, because this has become a part of our knowledge and memory of the twentieth century, but similar phenomena have happened even after that, in such varied places in the world as Tibet, Uganda, East Timor, and Cambodia, and, if we include less extensive mass murders, the list can become a long one. These things happened and can easily happen over and over again in the future whenever the efficient technology of killing is concentrated in the hands of one single individual who thinks in a very unusual manner. Certainly, merely because far more efficient technology was available to him, Hitler was much more effective and successful at killing than Nero or Ivan the Terrible.

MORE POWER TO EVERYONE

The history of technology shows that an innovation tends to spread from one source to larger and larger numbers of individuals. It can ultimately diffuse into all segments and subcultures within a society. Internationally, it can diffuse from one country to another. In a capitalist society, a consumer-oriented innovation may be expensive in the beginning, but, by means of mass production, it becomes less expensive over time.

As a result, the number of potential buyers increases significantly, and this in turn increases the quantity of production, and the new product becomes even less expensive, until it is within the purchasing ability of a significant portion of the population. Soon, less developed countries with less expensive labor costs begin to produce similar products, which are often inferior in quality but much cheaper, and these products are available to even wider segments of the population, until practically everyone can afford to buy one. Less developed countries themselves can market such products for domestic consumers, too. This phenomenon has been seen over and over again, and is especially true for electric and electronic products.

More or less similar statements can be applied to the diffusion of automobiles, motorcycles, motorboats, desktop computers, and many other products of recent years. Technological development also creates less sophisticated tools for manipulation, and, as the purchasing power of the consumers of the country increases, as is usually the case in a capitalist country most of the time, more and more tools are available for more and more people.

Some examples are sufficient to illustrate this point. The cassette headphone set is nowadays available to anyone, including children, and, as a result, it is quite common to see passengers in a bus or train with such a headphone set. They usually play rock music so loudly that others nearby are forced to hear the noise coming out of the headphone. This is certainly a form of noise pollution to those who do not want to be disturbed in this manner. Even worse is the stereo cassette player with loudspeakers. Again, this is a familiar source of noise pollution on downtown streets of large American cities.

Among the countries I have personally seen, however, the worst is Japan. In Japan, the extent of noise pollution is incredible, not because noise is made inadvertently in the functioning of society from such sources as factories, trains, airplanes, and automobiles, but because much of the noise is intentionally produced in order to manipulate people. For example, before an election, many campaigning cars with several loudspeakers move freely all over, running tapes of the messages from or about the party and the candidate. When two such cars from two different parties stay at a square, each presenting its version very loudly at the same time, neither side can be understood. It is an incredible scene suggesting a possible method of torture in hell.

Another form of torture is carried out by a fanatic religious sect. This sect also runs cars equipped with loudspeakers, releasing a recorded message at various places all over the country. If you do not believe in their religion, it is a sheer torture to be forced to hear about it. Indeed, the message tells that if you do not believe in their religion, you will go to hell, but you are in fact already in hell! Yet another form of noise pollution is caused by political activists in a similar manner, insisting on extreme political views, usually right-wing politics.

Noise pollution in Japan is not limited to these phenomena. Quite often, shopping malls, stores, trains, restaurants, elevators, and even libraries are equipped with loudspeaker systems, and one is forced to hear messages, advertisements, or music continually, whether one wants to hear it or not. Things are no better even when one is at home, because all kinds of peddlers nowadays come to residential areas driving cars equipped with loudspeakers. In a country like this, one is forced to see oneself as a miserable object for manipulation. Technology as the tool for manipulation has become so powerful that it destroys the dignity of and respect for the individual. The most amazing and frightening thing, however, is that Japanese people on the whole take this phenomenon for granted as a natural consequence of a "free" society; the expression "free" here of course means "freedom for" at the expense of "freedom from."

It is certainly possible to consider the case of Japan as an exception because there has never been a true tradition of liberalism or individualism in Japan, nor has there developed an indigenous Japanese philosophy of enlightenment. After all, most of the ideas and thoughts associated with industrial society, except capitalism, are alien to traditional Japanese culture. But I have seen similar phenomena in Western Europe and North America, and therefore these phenomena are certainly not peculiar to Japan as a non-Western country. Indeed, I was surprised and at the same time very disappointed when I first visited London in 1988. Many stores were as noisy as Japanese stores, playing recorded rock music very loud, even in music stores where some customers were there in order to buy classical records and tapes. I could not believe that this was the country where J. S. Mill once lived. Noisy vehicles vending in London reminded me of Japan.

If you like rock music, to be forced to hear it everywhere is not a problem at all. Rather, it would be a pleasure. Also, if you do not mind hearing noise of any kind, it does not matter, either. But if you hate rock music, as I do, this kind of experience is torture. This phenomenon occurs because electronic products have become so inexpensive that store owners and children can purchase them as tools for manipulating others or themselves. But due to the nature of the sound wave based on the vibration of air, everyone near the headphone or loudspeakers becomes the material to be worked upon indiscriminately. When this is combined with the phenomenon of psychological variation among us, some individuals are forced to suffer from it.

Parenthetically, I should add that totalitarianism is naturally against negative freedom. I vividly recall that, when I arrived at the central station in Moscow by train from Helsinki in 1971, I was at once forced to hear Mussorgsky's *Khovantchina* over the loudspeaker system at the train station. I like this particular work, but not in this manner! In a country like Vietnam, street loudspeakers are commonly seen, and become the tools for manipulation, with no "freedom from" allowed.

In addition to noise pollution, there is so-called graffiti on buildings, trains, buses, stations, and other public places, which has become another familiar phenomenon in many countries. This is no doubt a form of stimulus seeking behavior, mainly as play and creativity. Chimpanzees will do this just for pleasure.[14] Among our species, graffiti is known in archaeology and history. For example, ancient Greek mercenaries scribbled their names on an Egyptian sphinx, and a Greek from Pamphylia carved his name on the pyramid at Giza. The excavations at Pompeii revealed graffiti dealing with the gladiatorial shows.[15]

But, throughout much of human history, the means for scribbling (such as paint) have been limited. Above all, these means were not easily available to youngsters. But nowadays various kinds of pens, brushes, and spray paint based on chemically produced dyes are available to everyone, and youngsters often steal them at stores. Easy access to these materials means more graffiti everywhere.

A wide variety of tools of all kinds is also available not only to youngsters but also to political dissidents, pressure groups, interest organizations, religious, ethnic, and racial organizations, and other subcultural groups. The tools for terrorism, hijacking, or bombing are easily available to almost anyone nowadays, and therefore these phenomena are happening routinely all over the world.

MORE DIFFUSION BETWEEN CULTURES AND NATIONS

Archaeologists and anthropologists tell us that diffusion, or the spreading of objects and ideas from one culture to another, was much more extensive prehistorically and ethnologically than we might assume, and that this phenomenon has been almost worldwide. With the development of technology, this tendency has become even more extensive. The development of wide-body jets has made air travel accessible to a significant portion of the peoples in the world, and, as a result, as tourists, students, workers, businessmen, and refugees, they move much more easily from one nation to another and from one culture to another. The movement of peoples almost inevitably involves the diffusion of objects and ideas.

Another very significant innovation in stimulating diffusion is television, which can affect an even greater number of individuals in the world. Because it can let people know, much more effectively than other forms of mass media, how other people live in the rest of the world, television has no doubt helped a large number of individuals in the world decide to leave their own countries and cultures temporarily or permanently. Of course, most individuals stay where they are, but merely by being exposed to different cultures via television, the possibility for diffusion becomes significantly greater. New ideas easily diffuse in this way: a new pattern of behavior is imitated, or a material innovation is imported, smuggled, or copied locally.

Diffusion from the West into Eastern Europe, China, and other parts of the world—involving such phenomena as rock music, blue jeans, fast food, and other items of popular culture, as well as the magical word "democracy"—is one good example. Diffusion may also take place between subcultures within a larger culture. In the United States, phenomena that were earlier tied to special racial or ethnic groups have diffused into other segments of society, such as rock music, drugs, and various ethnic foods.

Greater contact between cultures and nations does not necessarily mean greater diffusion, but, in reality, with regard to popular culture and political ideology, there is a clear trend in the contemporary world in that direction. You are likely to hear or to be forced to hear rock music in Paris, Moscow, Tokyo, or Warsaw as much as in American cities. As far as some aspects of popular culture are concerned, the world is becoming homogeneous.

Theoretically speaking, homogenization of cultures in the world may help to reduce conflict between cultures. But this is possible only when there is room for those who do not like the homogenized world culture. Otherwise, the result is exactly like the tyranny of the majority in a "democratic" country. If one is forced to accept the tyranny of world culture, one may suffer from it. The nature of suffering in this situation lies not in differences between cultures but in psychological differences between individuals, and this is a phenomenon found in any culture. Thus, by making the cultures of the world homogeneous, you are replacing one form of conflict and suffering with another.

THE GREATER DIFFICULTY OF AVOIDANCE

As technology dominates our lives more and more, there is less and less practical possibility for evading the unwanted aspects of culture, for example, by becoming a hermit.

As I have discussed earlier, in agricultural society the psychological variations among people were not likely to be as great as they potentially could be, mainly because the physical and cultural variations in such a society were relatively limited, and because these variations did not actualize the full potential for psychological variation, which is to a significant extent dependent upon the physical and cultural variations to which one is exposed.

Yet even in this form of society there were often institutions for those who wanted to get away from the disturbances of mainstream culture. Typically, a certain religious or philosophical way of life was possible for such people. Monasteries and convents in the West and some Buddhist sects in the East allowed selected individuals to be alone and to escape most disturbances from the outside world.

In building a monastery, hostile attacks from the outside were assumed, and usually large piles of buildings were erected with strong outside walls, within which daily life could be maintained. Among many monasteries, the largest and most celebrated one was the abbey of Cluny in France. In the twelfth century, it housed a community of 400 individuals. But, unfortunately, it could not survive the rise of industrialization and the new way of looking at human beings stimulated by the philosophy of the Enlightenment. Between 1789 and 1814, during the revolutionary period in France, almost the whole complex of buildings disappeared.[16]

Monasteries could not escape being destroyed by communism, either. The monks in Russian monasteries were expelled after the revolution of 1917, and, if they were not worth preserving as national monuments of art, the buildings were destroyed. In the case of the Buddhist monasteries in India, it was partly the Moslem invasions that by about the thirteenth century, had destroyed them. In China, Buddhism and Taoism are being suspended by communism, including, naturally, their monasteries.[17]

Certainly, monasteries and convents still exist in the West, and even in the Soviet Union there may be some renewal of monastic life.[18] Yet life in such places now, in the industrialized world, is much more precarious than during the medieval period. Airplanes may fly over the monastery or convent, automobiles and motorcycles may be speeding down nearby roads, and the air, water, and land in the immediate environment may be polluted. Such institutions simply cannot escape the impact of technology.

Similarly, self-isolated communities such as the Amish, the Hutterites, the Dukhobors were possible in the past, but are now faced with increasing threats from the outside world. For example, during World Wars I and II, the Hutterites in Canada and the United States were targets of vandalism because of their pacifist convictions.[19] The Amish in Pennsylvania have been forced to educate their children more than they want, and their agriculture has been threatened by competition from the outside because, due to their religious beliefs, they refuse to use motor trucks and field tractors.[20]

It is clearly even more difficult for the individual who does not want to belong to a structured community with or without a religious basis. Furthermore, the global tendency for the pollution of air, water, and land, as well as noise pollution and the pollution of commercialism, allows nobody any means of escape.

THE EMERGENCE OF NEW SUBCULTURES

A culture beyond the level of a very simple hunter-gatherer type tends to have subcultures within it, although the actual differences between subcultures may vary considerably from one culture to the next. The existence of subcultures is an old phenomenon. But, as a result of technological development, industrial society is now characterized by new forms of subcultures as well. In many situations, subcultures of this new type are far more decisive in accentuating the cultural variations within society. This point can best be illustrated through a comparison of old and new forms of subcultures.

In the old form, subcultures are based mainly on such factors as racial, ethnic, linguistic, religious, and class differences. These differences were present already in advanced hunter-gatherer societies, and, in agricultural society, several of these subcultural differentiations within society were quite commonly found in various parts of the world.

But, after the industrial revolution, an additional factor creating subcultural differences emerged. This is the growth and differentiation of technological knowledge and skill *per se*. Certainly, even in agricultural society, a tendency for subcultural differences existed due to the specialized knowledge and skill needed by artisans. But the nature of their work was still comprehensible to most other individuals in the same society, and a person could more or less understand, or at least infer, what another person was doing in his occupation.

When the differentiation of knowledge and skill in technology was not extensive, a "jack of all trades" could exist and indeed often did exist. This was possible even in a considerably advanced state of cultural evolution. Leonardo da Vinci is a well-known example. Immediately before and after the industrial revolution, many significant achievements in science and technology were made by amateurs, some well-known examples being Antony van Leewenhoek, John Dalton, Michael Faraday, and James P. Joule.

In contemporary industrial society this is no longer possible. Not only that, nowadays scientists and engineers have difficulty in keeping up with the latest innovations and discoveries even within their own fields of competence, which are becoming more and more specialized every day. In view of the exponential growth of scientific and technological knowledge, on the one hand, and the finite amount of time each individual has for learning new information, on the other, it is clearly axiomatic that this is inevitable and will become more and more extensive in the future.

The technological specialist of this nature is what philosopher Ortega y Gasset calls the "technician." According to Ortega, the technician is competent to carry out a specialized task but is semi-ignorant of science, culture, and history. In politics, the arts, and the sciences not directly related to that speciality, the technician is primitive and ignorant. A person of this nature is self-satisfied and insensitive to things outside one specialized area of competence.[21] Ortega is critical of these "technicians" in the contemporary industrial West, but, unfortunately, this is an inevitable consequence of the exponential growth of scientific and technological knowledge.

The emergence and multiplication of specialist subcultures in industrial society are creating a new form of cultural variation. Whenever a new technological innovation becomes a reality, there is a very small number of specialists who have adequate knowledge about it, but, at the same time, the rest of the population is almost totally incompetent in evaluating it objectively. The division between specialists and nonspecialists applies, however, only to this one particular innovation; the division may be totally different with regard to another innovation. Since there are constantly new technological innovations, the number of divisions becomes greater and greater. Most individuals fall into the category of incompetent nonspecialists, belonging to the specialist category as to a very limited number of innovations, or not at all.

Despite their incompetence and lack of knowledge, most nonspecialists want to express their opinions on any innovation that is likely to affect everyone in society. By falsely believing that to let people express opinions toward an innovation that they are totally incompetent to evaluate objectively is "democratic" and therefore desirable, a "democratic" country often brings the issue to a public debate or referendum. However, many of these issues are controversial even among the specialists themselves to begin with, and to make such issues into topics of public debate does not guarantee the best solution for society.

Nuclear power plants, economic measures against inflation or stagnation, the pollution and destruction of the environment, genetic engineering, radiation of agricultural products, and many other familiar topics of public debate are beyond the competence of most individuals in society, and it may be meaningless or even harmful to let the incompetent masses intervene with the best solution arrived at by specialists. This is almost always true when measures against the deteriorating economy of a country are resisted by the masses either by a massive demonstration or by voting against them, when in reality the measures are the best practical solution in that particular situation.

Whatever the nature of the issue, when it is brought to the attention of the populace, a new form of cultural variation is created, including a small number of specialists and those who can understand their reasoning, on the one hand, and a large number of masses who lack specialized knowledge and the skills to understand the issue, on the other. Subcultural variation is further complicated in many cases by having divisions within both the specialist category and the category of the masses.

Additional complications may occur when new subcultural differentiation is combined with the old form of subcultures based on such factors as racial, ethnic, linguistic, religious, or class differences. A good example is the explosive controversy over the article by Arthur Jensen, in which the possibility for racial differences in intelligence was suggested. Since this article implied that Afro-Americans were genetically inferior to Euro-Americans in terms of intelligence, the content of the article was interpreted as "racism," arousing extensive protests and anger among Afro-Americans and "liberals." This is a very understandable reaction, but this article itself is highly technical, and most individuals are not competent enough to understand what Jensen is really saying.

In the past, Charles Darwin encountered a similar reaction, when most people could not really understand what Darwin was saying in his book. I am not saying that both Darwin and Jensen are right and that to reject their ideas is wrong. What I am saying is that in the age of specialists the masses cannot understand what a specialist says, yet they react violently to any disturbing idea, even when to consider such a new idea as a possibility is desirable for the sake of the advancement of knowledge.

THE DEATH OF LIBERALISM

As I stated earlier, technology as such is not necessarily against liberalism or negative freedom. Yet since it emerged and developed as a means for manipulation, it can be utilized in a variety of situations and in any political or social system. It is almost self-evident that in totalitarian countries such as Nazi Germany, dogmatic communist countries, or an Iran controlled by Islamic fundamentalists, political, religious, or ideological deviance is not tolerated, and technology is indeed an effective means of carrying out totalitarianism. In this situation, technology runs directly against liberalism.

It is often assumed that democracy is the only ideology opposed to totalitarianism. Indeed, the oppressed and persecuted individuals in totalitarian countries visualize democracy as the solution to their problem. To them, "democracy" is almost a pseudo-religious belief, and they are convinced that to change the totalitarian country into a democratic one will solve their problems. In a certain limited sense this may be true, if they are thinking of such very limited changes as the legalization of various political parties, free election, or freedom of the press. But beyond this, democracy may not be much better than totalitarianism.

The problem with democracy from the standpoint of liberalism lies in the tyranny of the majority, which has been so eloquently discussed by

Tocqueville. In democratic society, the belief in the majority has turned into a pseudo-religious conviction. According to this conviction, the masses are always right, and what they want and do is always right. Without such a false belief, indeed, it would be unthinkable to let the masses decide such issues as energy, abortion, or smoking in public places. This is the greatest mistake of democracy.

As J. S. Mill and Isaiah Berlin, among others, have pointed out, the idea of negative freedom has never been common even in the recent history of Western civilization. It is an idea of the minority. It is even alien to many. For this reason, the issue of negative freedom is not likely to become a matter of interest for the majority. In democracy, if the majority is not interested, that means death to the idea. When this problem is combined with commercialism, as is usually the case in democracy under capitalism, it is the masses that distinguish between acceptable and unacceptable ideas and actions. The ideas and actions that are accepted and liked by the masses dominate society. Of course, in reality, the masses may be manipulated by the interest for profit, as pointed out by many, such as Galbraith. When the masses are not interested in the biological fact and necessity for variation among us, and the importance of liberalism and negative freedom as a result, commercialism naturally takes the side of the masses. Commercialism always goes along with the masses, and they in turn help commercialism. This is one of the worst consequences of capitalism. In this way, liberalism, and, above all, negative freedom, is killed.

NOTES

1. Isaiah Berlin, "Two Concepts of Liberty," in idem, *Four Essays on Liberty* (London: Oxford University Press, 1969), p. 124.

2. Ibid. p. 131.

3. Ibid., p. 161.

4. Ibid., p. 132.

5. Ibid., p. 129.

6. Ibid.

7. Ibid., p. 161.

8. Ibid., p. 139.

9. Ibid., p. 171.

10. Ibid., p. 129.

11. John Kenneth Galbraith, *The New Industrial State* (New York: Mentor Book, 1967), pp. 212–13.

12. Bruce C. Netschert, "Chernobyl: The Sequence of Events," in *1987 Britannica Book of the Year* (Chicago: Encyclopaedia Britannica, 1987), p. 204.

13. Minoru Sano et al., eds., *Nijyusseiki Zen Kiroku* (The Compete Records of the Twentieth Century) (Tokyo: Kodansha, 1982), p. 1179.

14. D. Morris, *The Biology of Art* (London: Methuen, 1962).

15. "Graffito," *The New Encyclopaedia Britannica*, Vol. 5 (1987), p. 405.

16. "Monastery," *Encyclopaedia Britannica*, Vol. 15 (1969), pp. 684, 686.

17. "Monasticism," *Encyclopaedia Britannica*, Vol. 15 (1969), pp. 690, 694.

18. Ibid., p. 690.

19. "Utopia," *Encyclopaedia Britannica*, Vol. 22 (1969), p. 825.

20. John Gillin, *The Ways of Man* (New York: Appleton-Century-Crofts, 1948), pp. 209–20.

21. José Ortega y Gasset, *The Revolt of the Masses* (London: Unwin Books, 1963), pp. 82–87.

9

EVOLUTION AS TRAGEDY

To describe evolution as a tragedy certainly reflects my personal values. It is also possible to consider it as a comedy, because often tragedy and comedy are the same thing. It is tragic that we do not really understand what we have been doing and will be doing, but to live without knowing can be comical. In this sense, we might as well use the word "comedy" in describing human history, because we are perhaps experiencing totally unexpected consequences through biological and cultural evolution. A comedy is usually filled with unexpected events, and it is these unexpected events that are amusing and make us laugh.

Human history is certainly filled with such unexpected events. Whether or not these events are to be seen as sources of amusement or suffering is probably up to each individual. I personally see evolution as a tragedy mainly because of the nature and extent of the suffering brought about by it.

The problem of human suffering can be scrutinized by reconsidering our two important biological features: stimulus seeking behavior and the corresponding manipulatory drive; and physical, psychological, and cultural variations. Since much of our suffering results from one or both of these biological phenomena, it is theoretically conceivable that to change one or both of these aspects of our background may reduce our suffering. Since both are biologically given, so to speak, we are here dealing with the possibility for intervening with them in one way or another.

INTERVENTION IN THE PROCESSES OF THE MANIPULATORY DRIVE

First, theoretically, to reduce the exercise of the manipulatory drive, and, as a result, to reduce the extent of manipulation, is a conceivable possibility. The reduction of manipulatory activities can mean the reduction of conflict in all forms, including war, genocide, racial, ethnic, religious, and linguistic conflicts as well as homicide, abuse of women and children, and intergenerational conflict at the more personal level.

Here I am not talking about attempts to change our attitude to make us more peaceful and less hostile to each other. Unfortunately, human history clearly shows that the manipulatory drive in us is so strong that to try to reduce or prevent it psychologically by appealing to reason, logic, or rationality has repeatedly turned out to be very ineffective. Educating or conditioning people may have a relatively minor effect in this regard.

Certainly, there are cultures that have been described as peaceful. For example, the Semai of Malaya were once described as a peaceful people. Children were not struck and were raised in the spirit of nonviolence. Murder was unknown in this culture. When Semai men were recruited by the British colonial government to fight communist guerrillas in the early 1950s, many outsiders thought such a peaceful people would not be able to fight. Above all, they did not know that soldiers were to fight and kill the enemy. But when communist terrorists killed the kinsmen of some of the Semai soldiers, they suddenly became unbelievably aggressive. A Semai man described what they had done: "We killed, killed, killed. . . . We thought only of killing."[1]

The moral of this story is that, even when a person is trained and conditioned to be peaceful and nonaggressive since childhood, aggression can be quite easily activated and released as much as in anyone from an aggressive culture. In this sense, aggression is much stronger than culture in directing human behavior.[2]

Restriction of the manipulatory drive in other forms has been repeatedly shown to be ineffective. Everyone knows that to drive a car pollutes the air. But this behavior is so rewarding in satisfying the manipulatory drive that various attempts to reduce the maximum speed limit or the total amount of automobile traffic are consistently unsuccessful. Similarly, if we look at human history as a whole, we continue to produce weapons that are more and more effective, and we now have reached the point where we can easily exterminate ourselves.

What can we think of, then, as a means to reduce our excessively strong manipulatory drive? In the near future, it may be possible to reduce the intensity of the manipulatory drive by means of genetic engineering. The international project called HUGO is now preparing the complete mapping of the human DNA, and when this is ready, it is quite conceivable to locate relevant arrangements of the heterocyclic amines associated with the manipulatory drive as well as its intensity. Certainly, there will be individuals in the future who suggest this as a measure to prevent us from self-extermination—if we have not exterminated ourselves before that. But the phenomenon of human variation also suggests that there will be those who violently reject such a measure. Exactly as in the debates over environmental pollution, danger of thermonuclear war, and application of genetic engineering to plants and animals, the idea of manipulating ourselves will create an emotionally charged debate.

On the one hand, there are individuals who want to apply their manipulatory drive even to our DNA in order to reduce the harmful effect of manipulatory drives that are too strong, and, on the other, there are individuals who do not want to manipulate human DNA because they want the present state of our manipulatory drive to continue. There will certainly be many other viewpoints in this issue, with the arguments based on various forms of religious, political, cultural, or philosophical ideas, or simply ignorance.

The outcome of such a debate in the future is difficult to predict, because the feasibility of carrying out such a plan is analogous to the practicality of disarmament talks, and because the form of political system can be a factor as well. A strongly totalitarian government as well as the tyranny of the majority in a democracy can carry out totally unexpected things. Yet it is clear that the genetic control of the manipulatory drive is not possible now. In the meantime, the manipulatory drive makes us produce more and more technological innovations to manipulate material and symbolic objects effectively, and this makes us suffer.

THE REDUCTION OF HUMAN VARIATION

Another possibility for reducing our suffering is to reduce our physical, psychological, and cultural variations. As in the case of the manipulatory drive, our variations are difficult to control. Psychological and cultural variations are reduced to some extent in our daily lives by fiat at the expense of extreme individuals on both sides of the bell-shaped distribution curve. In the ideology of democracy, it is assumed that the middle range of opinions offers the most realistic solution whenever opinions vary as to a certain issue, as is usually the case.

In the totalitarian system, the opinion of those in power is enforced upon the masses even when their opinions are significantly or radically different from those of the masses. Whichever is the case, different opinions are, in effect, ignored. The most ironic point here is that in order to satisfy some individuals in society and to allow them to realize their manipulatory drive, others are forced to refrain from using their manipulatory drive.

This method has been known in human history for a long time. Morality, ethics, "commonsense," the law, folkways, social pressure, and religion are all customary means of dealing with this problem and reducing human variation in society. In the contemporary world, international law, international customary practice, and international pressure, which are mostly derived from the law and commonsense of the modern West as well as Christianity, function similarly to regulate this among the nations and cultures of the world.

At the more personal level, psychotherapy, counseling, and education are the means to deal with psychologically extreme individuals. Phobic persons, homosexuals, and other psychologically unusual individuals are consulted or treated so that they may become more like most other individuals in society. The problem with this approach in dealing with psychological and cultural variation is that it almost always makes extreme individuals suffer, because their uniqueness is ignored. This is not at all the solution to the problem as I see it.

It is most likely that, in the future, there will be an attempt to apply genetic engineering in this situation, too, as a conceivable solution to human variation. Cloning, for example, can solve this problem neatly and once and for all by making everyone alike physically. By eliminating the physical variation among us except for the sexual difference, psychological variations can be reduced significantly as well. After that, cultural variation can also be reduced significantly.

But the phenomenon of human variation is to be seen as an advantage rather than as a burden for our survival. Variations are resources for adapting to the changed conditions in the environment. To make all individuals of *Homo sapiens sapiens* homogeneous can mean an extremely limited range for adaptation, and a slight change in the environment may result in a disastrous inability to adapt physically, psychologically, and culturally. It is therefore very unwise to try to reduce our variation. Exactly as in the case of the manipulatory drive, we must accept human variation as a condition in life. It must not be eliminated. But, at the same time, as we have seen, the phenomenon of variation itself creates human suffering, such as prejudice and discrimination, and conflict of all kinds in combination with the manipulatory drive, as we have discussed earlier.

Here is a dilemma in which neither of the two alternatives is desirable. First, cloning or similar methods in genetic engineering can reduce or eliminate human variation, and, as a consequence, all problems resulting from human variation can be reduced. But this may result in self-extermination. Second, to continue to have human variation is to perpetuate human suffering. Certainly, the feasibility of applying cloning or a similar technique of genetic engineering to ourselves is not yet in sight, but that does not matter. If we accept the survival of ourselves as an axiomatic postulate and requirement, to intervene in human variation is not the method to deal with human suffering.

To try to reduce the intensity of the manipulatory drive by means of conditioning, culture, or education is ineffective, as human history thus far clearly indicates. Genetic engineering can be an effective method in the future in this regard, but this will become a most controversial issue, in view of our physical, psychological, and cultural variations. The conclusion we derive by reconsidering the two biological aspects of ourselves is that at least until we can reach a realistic solution, neither the manipulatory drive nor variation can be intervened with effectively in reducing human suffering. This is a source of human tragedy.

HUMAN SUFFERING IN THE PAST

Certainly, humans suffered in the past, too. But the suffering we have now is different from the older form of suffering both in nature and in the kinds of individuals affected. These changes in suffering can be clarified if we look at suffering in hunter-gatherer society, agricultural society, and industrial society.

In hunter-gatherer society, as you will recall, physical, psychological, and cultural variations are minimal. Racially and ethnically, people are homogeneous, and their culture is also homogeneous, in the sense that obvious subcultures do not exist. As a result, people are also psychologically relatively homogeneous without physical and cultural variations. Genetically, they are adequate enough to actualize psychological variation as much as we do now, but the catalyst of physical and cultural variations is absent.

People in hunter-gatherer society, then, are relatively free from the suffering that results from physical, psychological, and cultural variations. The only exception is the possibility for contact with another tribe or culture, which can become a source of conflict and war. But these phenomena are often similar to conflict between families, lineages, or clans, and they are by no means comparable to the mass killing that takes place in modern war.[3]

With relatively little variations in the physical, psychological, and cultural dimensions, the predominant form of suffering in hunter-gatherer society is caused by the elements. Extreme cold and heat, flood, earthquake, fire, volcanic explosion, and drought make people suffer directly and also indirectly, by causing famine and destroying their game. Illness is also likely to affect people more when they have less knowledge about the causes of illness and the ways for treatment and prevention.

In this situation, suffering in hunter-gatherer society may be said to have three characteristics. First, the causes of suffering are external to people. Since the elements, natural catastrophe, and the causes of illness basically lie outside the biological and sociocultural spheres of people and are external to them, the sources of suffering are mostly beyond these people's control. In a sense, suffering is forced upon them externally, and, with their limited technology, they must accept it.

Second, the nature of suffering is objective and material. When a person suffers from extreme heat or cold, famine, an illness, or a natural catastrophe, he or she is suffering from a condition that is easily recognizable objectively, such as shivering, sweating, hunger, or high fever. These conditions can be dealt with through material measures to relieve the suffering. Clothes, shelter, food, water, or medicine can be supplied to such a person whenever available, and this form of suffering can be reduced or eliminated relatively easily.

Third, these causes of suffering are likely, more or less, to affect all individuals in society. The elements certainly affect everyone, and so does a natural catastrophe. With relatively little division of labor and few class differences, illness is also likely to affect all individuals in society equally. This would certainly be true for epidemics.

After the neolithic revolution, the possibility for physical, psychological, and cultural variation increased. The emergence of differences in social class, occupation, or religion meant the possibility for the development of subcultures within society. The stratification of society was in some cases accompanied by and based on racial or ethnic differences. All these changes paved the way for actualizing the potential for the psychological differences that people genetically acquired. In this way, suffering due to physical, psychological, and cultural differences began to appear in agricultural society. This means that the three characteristics of suffering in hunter-gatherer society began to change.

First, the causes of suffering became less external. Compared with hunter-gatherer society, and thanks to the advancement of technology, agricultural society is better at reducing the impact of the elements upon people. In this sense, the external sources of suffering are reduced in intensity, and people suffer less from external sources. But this does not mean a reduction in suffering, because there are new forms of suffering due to physical, psychological, and cultural variation, and these sources are not external to people; they are internal and, so to speak, built into them.

Second, the nature of suffering became less objective and less material. A person may now suffer not so much from the cold, but from the way others treat him or her because of his or her physical appearance, occupation, or way of thinking. By being different from others physically, culturally, or psychologically, a person may be forced to suffer, and this cannot easily be reduced by material means, because the sources of suffering lie in other human beings rather than in ecological or geographical reality. Furthermore, the intensity of suffering is subjective, because it is caused by the way a person experiences how others treat him or her.

Third, these causes of suffering do not affect everyone in agricultural society equally. Some individuals in society may suffer because they are different by belonging to a certain ethnic or racial group, or to a religious or occupational group, or by psychologically responding to these phenomena in society in a different way.

Of course, these differences in the nature of suffering between hunter-gatherer society and agricultural society are a matter of degree; all I would like to say is that, compared with hunter-gatherer society, agricultural society is characterized by a tendency to move away from the external nature of suffering, which is objective and material, affecting all individuals in society as typically found in hunter-gatherer society. Depending on the complexity of agricultural society and its closeness to industrialization, these changes in the nature of suffering vary considerably. On the whole, agricultural society may be visualized as a transitional form with regard to the nature of suffering.

HUMAN SUFFERING AFTER INDUSTRIALIZATION

A truly drastic change was triggered by the industrial revolution, and, in contemporary industrial society, the nature of suffering is clearly seen in conditions almost opposite to those found in hunter-gatherer society. Indeed, these changes parallel the contrast repeatedly discussed by sociologists and anthropologists in the comparison of agricultural and industrial societies. But these sociologists and anthropologists have not shown much interest in changes in the nature of suffering in connection with the transition from agricultural society to industrial society. Conceivably, suffering of the individual in society has never attracted their attention. For this reason, it is perhaps desirable to explain how the nature of suffering in industrial society differs from that in hunter-gatherer society.

Industrial society is characterized by the extensive actualization of the two biological features built into us: the strong manipulatory drive is expressed in dominance of science and technology in society; and physical, psychological, and cultural variations have been brought about by technology. Technology has directly created physical and cultural variations by bringing peoples into closer contact, which in turn actualized their potential for psychological variation. As a result, industrial society is characterized by three new forms of suffering.

First, the causes of suffering lie mostly internal to us. Certainly, we cannot prevent earthquakes, typhoons, droughts, or many kinds of illnesses, but we have accumulated more knowledge about these sources of suffering, which helps us to avoid them or to reduce their impact on us to a certain extent. Above all, suffering caused by the elements has been reduced or eliminated drastically by means of technology. But, at the same time, due to greater variations in the physical, psychological, and cultural dimensions in society, an individual is now faced with new sources of suffering. These variations are within us rather than external to us. These three forms of variation create conflict among us, and the sources of a person's suffering lie in other human beings.

Second, the nature of suffering is subjective and less material. The individual in industrial society suffers from prejudice, discrimination, alienation, depression, unemployment, failure in the struggle for power or success, and many other problems of the complex industrialized environment rather than from the harshness of the elements. It is more difficult to understand these forms of suffering than suffering from heat, cold, or hunger. It is difficult to understand how it feels to be seen negatively and discriminated against without being in such a situation. It is easy to sympathize with a person suffering from a famine, but it is difficult to understand how an alienated person feels. Suffering in these forms is subjective and harder to deal with by material measures. Technology cannot easily be applied to these forms of suffering.

Third, suffering in industrial society usually does not affect everyone in society equally. When the source of suffering is mostly internal rather than external to us, people suffer from physical, psychological, and cultural variation, and this does not affect everyone. This is exactly why and how the idea of negative freedom emerged and has been ignored.

Psychological variation means that, in response to the same objective stimulus, people may vary in the extent to which they are affected or disturbed. A good example is noise pollution. It is usually only one segment of the population of any industrial country who feels noise in the environment as disturbing and irritating. Even when people are exposed to issues dealing with a matter of life or death, such as pollution, thermonuclear weapons, destruction of the ozone layer, and the development of the greenhouse effect, it is only a minority of people who take the matter seriously. Only a segment of this minority suffer from knowing about these problems and also from being unable to do anything about them.

Similarly, the physical variation of people and peoples can be a source of suffering only for those who are negatively affected by being different from others. The rest of the population in society or in the world does not suffer. In the case of cultural variation, those within the disadvantaged subculture may be additionally exposed to the external source of suffering by having, for example, poor housing and sanitary conditions. Their suffering can be further intensified by physical and psychological variations. But these forms of suffering may be totally alien to others.

Thus, the nature of suffering has changed its character through the evolution of human culture. To repeat, in hunter-gatherer society, the sources of suffering are external to people. They are objective and material, and suffering affects almost everyone equally. But in industrial society, the causes of suffering lie mainly within us. The nature of suffering is subjective and less material, and only some of the people in society suffer. These features of suffering are almost the exact opposite of those in hunter-gatherer society. The people who suffer in industrial society are those who find themselves significantly different from the majority in one way or another, and they suffer as a result of the advancement of technology, which has been instrumental in increasing the physical, psychological, and cultural variations in society. Here is a head-on collision between the strong manipulatory drive, on the one hand, and significant variation among us, on the other. The victims are the products of the incompatibility between the two biological phenomena that we have genetically acquired.

It is quite ironic that the strong manipulatory drive via technology has created fully actualized variations among us. Varied ideas, views, philosophies, religions, and political ideologies are available as a result. Variations are an advantage for a living species, and they should help us survive better, because different ideas are valuable with a great potential for contributing to our survival. But in reality, to be different quite often means punishment. Contribution to the species by being varied is dealt with by punishment. This is ironic and at the same time tragic.

VARIOUS ASPECTS OF THIS TRAGEDY

To describe the history of biological and cultural evolution as a tragedy may be vague and unclear, and it is probably desirable to state in what ways it has turned out to be a tragedy.

To have a strong manipulatory drive *per se* may not be tragic. Living organisms, especially evolutionarily advanced ones, tend to have a strong manipulatory drive. When a lion kills the young of a lioness fathered by another lion before mating, it is most likely that the lion does not suffer from moral or ethical considerations. These issues are totally irrelevant to and absent from a lion's life. The young that are being killed by a lion may experience fear before they die completely, but it is not clear to what extent they think about the consequence of their death, life after death, and so on. It is most likely that they do not know that among them, the young may be killed by an adult lion other than their own father. In this sense, their awareness of being killed lasts only a brief moment, and this potentially applies to all of the young. The experience of this nature is quite common in the organic world, and this is not sufficient to be considered as a tragedy. This is more or less a part of life.

But when the manipulatory drive is realized to excess, there emerges the danger of tragedy. First, in the case of *Homo sapiens sapiens*, physical strength itself is not sufficient to make its manipulatory behavior tragic, but rather it is the evolution of more and more powerful technology that makes it so, certainly when human technology may possibly exterminate not only the human species, but other living species as well. The phenomenon of one species becoming extinct itself is one of the most common and familiar stories in the organic world, but to exterminate itself is another matter. Especially when self-extermination is a consequence of biological and cultural evolution, it is tragic. One begins to wonder and to question the reason, meaning, and purpose of evolution itself.

Second, the fact that we must live with the impossible task of living as a species with an effective and very strong manipulatory drive, on the one hand, and physical, psychological, and cultural variations, on the other, is tragic. These two aspects of our biological heritage repeatedly turned out to be incompatible. Yet, as a biological species, we must live with this.

Third, if we did not know about this incompatibility, its tragic nature would be less painful. We would be born and die just like most other species. But in reality, we all know this quite well. Throughout human history, we have produced novels, fine arts, music, and theatrical works dealing with exactly this point. We know this, but we cannot do anything to change it. This is a common theme in many countries and cultures throughout human history.

Fourth, this incompatibility has more or less always existed in human history, but, along with the evolution of technology, the incompatibility has become more and more intense and urgent; yet we still cannot do very much about it.

Fifth, extreme individuals in a given population or category of human beings are biologically necessary, and, by being different from the majority, they contribute to the species as a whole by being "banks" for physical, psychological, and cultural variations, which may be useful under different conditions of life. Yet they are more likely to be ignored, rejected, persecuted, or even killed by merely being different from the majority. Especially in the contemporary world of Western democracy, in which the "tyranny of the majority" behaves like a dictator, extreme individuals are quite often unjustly treated under the magic name of "democracy." The death of liberalism, especially in the form of giving up the idea of negative freedom as described by Isaiah Berlin, symbolizes this in the presumably "enlightened" world of Western democracy. Western democracy, as a product of the philosophy of the Enlightenment, is crushing and killing liberalism. Variation has become a source of suffering.

Sixth, as a biological species, variation is necessary. Yet we have attempted and carried out genocide in history. In the future, cloning of human beings can become an issue for debate and may even be carried out, if we remember the extreme psychological variation of our species.

Seventh, both liberalism and individualism emerged as products of the philosophy of the Enlightenment. This system of philosophy more or less concurrently emerged along with the industrial revolution, capitalism, and the gradual emergence of the masses. However, ironically and tragically enough, liberalism and individualism, both of which are most realistic ideas in view of human variation, have been crushed by them. Technology created only "freedom for" at the expense of "freedom from."

These are the main points of this book, which can be seen as the reasons for considering evolution as tragedy.

EVOLUTION AND CIVILIZATION

We tend to assume that biological evolution is desirable and good. When we discuss evolution, we think of better adaptation to the environment through time, such as the improvement of the sensory organs, the development of warm-blooded mammals, a more refined and effective reproductive system, or the improvement of the brain. As in the case of evolutionism as a philosophical thought, evolutionism as a natural scientific thought was a product of the philosophy of the Enlightenment, and, as a result, it is characterized by the assumption of more complexity, improvement, refinement, specialization, and, above all, progress over time. Evolution is a good and desirable thing, and it is a blessing to us because we are presumably the most evolved species.

At the same time, aside from the optimism of evolutionism, there is a perspective in the philosophy of history in which civilization is visualized as something similar to an organism. According to this persrpective, a civilization is born, matures, reaches its height, declines, and then dies, just like an organism. Among several advocates of this perspective, no doubt the best example is Oswald Spengler, whose major work *The Decline of the West* became a bestseller immediately after the chaos and confusion following World War I.[4]

Spengler has made us think by calling our attention to the aspect of decline and death of civilization and by popularizing the organic view of civilization. His book is a great intellectual achievement, but in my opinion, the weakness of his idea lies in conceiving of a whole civilization as an organism. To him, it is a civilization that is born and dies. Certainly, we may say so figuratively, but empirically this is not only misleading but also incorrect. Spengler was highly influenced by the Romanticism of the nineteenth century, in which such concepts as *Geist* (spirit), *Volk* (people), and *Gestalt* (form), were very important in understanding culture and civilization. Probably the best example is found in Hegel's works, and Spengler presented his view on civilization within the intellectual climate of Germany, which shaped and stimulated his thought.

Although both the biological theory of evolutionism and the idea of civilization as organism deal with the organism, surprisingly enough, little attempt has been made to consider these two forms of thought together. A conceivable reason is that evolutionism is optimistic and the idea of civilization is pessimistic. Another reason is probably that Spengler did not state the processes of birth and death of civilization in a way compatible with the scientific perspective of biological evolution.

If we follow the perspective presented in this book, however, it is possible to look at the phenomenon of civilization in terms of biological evolutionism. Instead of looking at the whole civilization as the unit and focus of interest, we can study the same problem by examining the individual as the ultimate agent for creating a civilization. If we borrow medieval philosophical terms and modify them somewhat, instead of

Spenger's "realistic" approach, another conceivable approach is a "nominalistic" one, shifting the focus of interest from a civilization as a totality to the individual as the creator of civilization.

If we take this "nominalistic" approach, then, the rise and fall of civilization can be attributed to the way in which the individual in society thinks and behaves. I assume that the manipulatory drive is a biological given for all human beings. Although there may be significant variations from one individual to another within a given population, the average intensity, if we measure it in one way or another, would be very much the same from one group to another, whether the basis of categorizing individuals be ethnicity, race, religion, culture, society, or nation.

It is logically conceivable that, when the manipulatory drive is expressed constructively in the manipulation of material and symbolic objects, a possibility for creating a civilization is present. When physical, psychological, and cultural variations of the individuals in such a unit are utilized effectively without being prevented by the variations themselves in expressing the manipulatory drive, a rich and creative civilization can be created. When the possibilities for expressing the manipulatory drive are exhausted, the civilization becomes stale, vulgar, repetitive, redundant, and unstimulating. The civilization has lost its creativity.

When a civilization reaches this stage, individuals attempt to express their manipulatory drive in new ways, but the results are neither productive nor cumulative. The civilization thus loses the condition for its existence, declines, and disappears. This is a conceivable scenario of a civilization if we take the two biological phenomena of the manipulatory drive and variation into account. In this sense, the conditions for a civilization are (1) skillful and constructive application of the manipulatory drive to the environment, and (2) wise utilization of human variation. As soon as a civilization fails to maintain these two conditions, it is bound to decline and "die" in the near future.

THE FATE OF WESTERN CIVILIZATION

Western civilization is no doubt one of the most impressive civilizations in human history. If we look at the sheer quantitative aspects, such as the total number of individuals involved, the extent of influencing and changing the environment, and the total number of artistic products, no doubt it is the "greatest" civilization we have ever had on the face of earth. It has become almost worldwide.

The secret for the success of Western civilization thus far certainly can be understood in terms of the two conditions I have suggested. Technology as the key tool in manipulating material and symbolic objects has become consistently better and more effective, showing an exponential improvement through time. The ideologies of democracy and equality have encouraged and realized the acceptance and utilization of physical, psychological, and cultural variations among us to a considerable extent. These changes are products of the philosophy of the Enlightenment. Indeed, this philosophy has made Western civilization unique and has helped it to excel in relation to other civilizations.

But what about the future of Western civilization? Is it dying, as Spengler told us? I think the crisis of Western civilization can be located in the two conditions for a civilization. The excessive emphasis on technology in Western civilization has been a great advantage in making it a truly great one when it is applied to material and symbolic objects constructively, so that the results can become cumulative. But when technology is applied otherwise, there is a serious danger. We all know many examples, such as inane programs on television, video films of very poor quality, excessive commercialism and mass advertisement based on high technology, production of synthetic narcotics, thermonuclear weapons, popular music of sheer noise, and "fine arts" emphasizing only quantity. This factor is especially serious because Western civilization has produced the most efficient and effective technology in human history. Its destructive use is disastrous.

The other condition for a civilization is not truly satisfactory, either. In order to create and maintain a civilization, it is necessary to take advantage of the physical, psychological, and cultural variations of people. By doing this, we can derive varied ideas for manipulating material and symbolic objects constructively. To deny variation is to kill the civilization. Yet there have been such attempts in Western civilization, and these developments can become contributing factors in killing it. We have seen the idea of "equality" emphasizing the equality of outcomes. We have seen the phenomenon of giving up negative freedom. To ignore human variation is to ignore our biological heritage, and this is totally unrealistic and, at the same time, a threat to Western civilization.

Compared with earlier civilizations, Western civilization is unique in two ways. First, it has developed technology far more significantly than any other civilization. Indeed, it has become so powerful that almost every individual of *Homo sapiens sapiens*, whether or not he or she is a member of Western civilization, is under its influence in one way or another. Second, in the contemporary world, the dominance of Western civilization has established itself on the whole globe to such an extent that the fate of Western civilization has become almost the same as the fate of *Homo sapiens sapiens*. We cannot talk about the future of our species without talking about Western civilization. Thus, what I consider as the tragedy of evolution has been basically brought about by Western civilization through its highly powerful and efficient technology.

The possible decline and extinction of Western civilization has made many of us worry about the future of our species. There have been suggestions and speculations about the future as well as about "human nature." For example, some optimistically believe that we are intelligent enough not to exterminate ourselves by means of a worldwide thermonuclear war, extensive pollution of the environment, or the destruction of the ozone layer. Others say that they "believe in people." Still others say that we are "rational enough to plan for the future" or "rational enough to understand the stupidity of exterminating ourselves," and so on.

These ideas may be comforting when we think about the possible danger of extinction. But they are, biologically speaking, not realistic enough, because their views are based on the mistake of categorical thinking. Certainly, some people can be intelligent, rational, or good enough to be trusted, but in view of the wide variation among us, there are also individuals who are not like that. As more and more powerful and efficient technology becomes available to more and more individuals, these self-comforting ideas become less and less realistic.

To try to look at the matter in terms of cultural relativism does not solve the problem either. According to a point of view maintained by some philosophers and anthropologists, there is no absolute way of evaluating cultures. This perspective was at one time rather popular among anthropologists, who began to discover the extensive variations throughout cultures, and were amazed by them.[5] But the experience of the Nazi Holocaust has made most of us reconsider the value of cultural relativism. It may have some academic value as a focus for intellectual analysis, but in practice it is not only useless but also dangerous. No matter what one feels and thinks, the threat and danger of technology is real, and to accept cultural relativism means the acceptance of unlimited application of technology upon ourselves, including genocide and the extermination of *Homo sapiens sapiens* completely. The danger of accepting all cultures as different yet valid is evident.

Certainly, there can be other ways of looking at the problem. For example, it may be possible to present a totally different picture of ourselves and our future in terms of, for example, theology or mysticism. I do not do that in this book merely because I am totally incompetent to do so. Indeed, if I follow one of the two postulates that I assume to be axiomatic, there should and will be many different ways of looking at the same problem. It is possible to argue that to ignore the phenomenon of human variation is desirable and good, or that to have the strong manipulatory drive is a blessing. For the same reason, I naturally expect that there are individuals who think what I say in this book is absurd and without any value whatsoever, but, at the same time, I also expect that there are individuals who agree with me and accept my ideas. I have written this book for both kinds of individuals as well as for those who react to my ideas in still different ways.

NOTES

1. Robert K. Dentan, *The Semai: A Nonviolent People of Malaya* (New York: Holt, Rinehart and Winston, 1968), pp. 58–59.

2. Konrad Lorenz, *On Aggression* (New York: Harcourt, Brace and World, 1966).

3. Keith F. Otterbein, "The Anthropology of War," in *Handbook of Social and Cultural Anthropology*, ed. John J. Honigmann (Chicago: Rand McNally, 1973), pp. 944–47.

4. Oswald Spengler, *The Decline of the West*, trans. Charles Francis Atkinson, 2 vols. (New York: Alfred A. Knopf, 1926, 1928).

5. Melville J. Herskovits, *Man and His Works* (New York: Alfred A. Knopf, 1952), p. 76.

GLOSSARY

allele Any of several genes determining alternative characteristics such as eye color. Alleles occupy the same locus on two homologous chromosomes.

bipedalism A mode of locomotion using two hind legs.

brachiation A mode of locomotion among some primate species in which an animal swings from one hold to another by using its arms.

cloning A method of producing one or more genetically identical organisms asexually from a single ancestor.

diploid The condition of having the total number of chromosomes.

ethology The scientific study of animal behavior.

gamete Reproductive cell, namely ovum or sperm.

genotype The genetic arrangement of an individual.

haploid The state of having only half the full count of chromosomes.

heterocyclic amine A chemical compound that functions as a unit in forming genetic information.

heterozygous The condition in which different alleles exist at a given locus on two homologous chromosomes.

home range A fixed area that is occupied by one or more animals.

homozygous A state of having the same allele at a given locus on two homologous chromosomes.

identification The process and result of placing oneself in another's position through imagination and understanding.

identification with the aggressor The process and result of placing oneself in a powerful, threatening person's position through imagination and understanding, resulting in the vicarious experience of power and the reduction of the sensations of threat and fear.

learned helplessness   An animal learns that it cannot escape from an unpleasant state and is helpless.

manipulatory area   A physical space consisting of objects that an organism sees and is able to manipulate.

material self   The material objects that one thinks are one's own, including one's body, clothes, home, and country.

meiosis   A process of cell division in which the number of chromosomes in the cell is halved.

mitosis   A process of cell division into two identical cells, both of which carry identical chromosomes.

negative freedom   The absence of coercion by others.

neolithic revolution   A totally new way of life in human history, characterized by the domestication of animals and the cultivation of plants.

nominalism   A philosophical position that maintains that abstract things and general ideas are mere names and have no independent reality. In social psychology, nominalism refers to the approach that emphasizes the individual, rejecting such concepts as culture, society, and civilization.

phylum Chordata   A taxonomic division of the animal kingdom including all vertebrates and certain aquatic organisms.

phynotype   The observable characteristics of an individual.

pluralism   A viewpoint assuming the existence of diverse political ideas within society.

realism   In medieval thought, realism, in opposition to nominalism, refers to the doctrine that universals have a real, objective existence. In sociology and anthropology, E. Durkheim, A. L. Kroeber, and L. A. White take an analogous position.

recessive   A condition in which an allele is dormant in combination with another allele.

relative deprivation   Awareness of being deprived in comparison with others.

sociobiology   The scientific study of animal behavior explicitly taking evolutionary factors into account.

status incongruence   The condition of having two or more statuses so that a person is not evaluated consistently, such as high income and low occupation.

territory   A specified area occupied and defended by one or more animals.

total institution   An institution in which all aspects of a person's life are planned and regulated in a group setting under a single authority.

SELECTED BIBLIOGRAPHY

Arblaster, Anthony. *The Rise and Decline of Western Liberalism*. Oxford: Basil Black-well, 1984.

Baldwin, J. D., and J. I. Baldwin. *Beyond Sociobiology*. New York: Elsevier, 1981.

Baum, Andrew, and Jerome E. Singer, eds. *Application of Personal Control*. Hillsdale, N.J.: Lawrence Erlbaum Associates, 1980.

Bell, Daniel. "The Public Household—On 'Fiscal Sociology' and the Liberal Society." *The Public Interest*, 37 (1974), pp. 29–68.

Bentley, Arthur. *The Process of Government: A Study of Social Pressures*. Chicago: University of Chicago Press, 1908.

Berlin, Isaiah. *Four Essays on Liberty*. London: Oxford University Press, 1969.

Brehm, J. W. *A Theory of Psychological Reactance*. New York: Academic Press, 1966.

Brinton, C. *The Anatomy of Revolution*. New York: Vintage, 1938.

Charlton, W., ed. *Aristotle's Physics, Books I and II*. Oxford: Clarendon Press, 1985.

Ciochon, R. L., and R. S. Corruccini, eds. *New Interpretations of Ape and Human Ancestry*. New York: Plenum Press, 1983.

Ciochon, R. L., and John G. Fleagle, eds. *Primate Evolution and Human Origins*. Menlo Park, Calif.: Benjamin Cumming Publishing Co., 1985.

Coleman, James S., et al. *Equality of Educational Opportunity*. Washington, D.C.: U.S. Government Printing Office, 1966.

Constant, Benjamin. *Political Writings*. Translated and edited by Biancamaria Fontana. Cambridge: Cambridge University Press, 1988.

Dentan, Robert K. *The Semai: A Nonviolent People of Malaya*. New York: Holt, Rinehart and Winston, 1968.

Edelman, Gerald M. *Neural Darwinism: The Theory of Neuronal Group Selection*. New York: Basic Books, 1987.

Erikson, Erik H. *Childhood and Society*, 2d ed. New York: W. W. Norton, 1963.

Fagen, R. *Animal Play Behavior*. New York: Oxford University Press, 1981.

Foster, George M. "Pleasant Society and the Image of Limited Good." *American Anthropologist*, 67 (1965), pp. 293–315.

Freud, Anna. *The Ego and the Mechanisms of Defense*, revised edition. New York: International Universities Press, 1982.

Goffman, Erving. *Asylums*. Garden City, N.Y.: Doubleday, 1961.

Goodall, Jane. *The Chimpanzees of Gombe: Patterns of Behavior*. Cambridge, Mass.: Belknap Press, 1986.

Harlow, H., M. K. Harlow, and D. R. Meyer. "Learning Motivated by a Manipulatory Drive." *Journal of Experimental Psychology*, 50 (1950), pp. 228–34.

Hitching, Francis. *The Neck of the Giraffe: Darwin, Evolution, and the New Biology*. New York: New American Library, 1982.

James, William. *The Principles of Psychology*, vol. 1. New York: Dover Publications, 1950, chap. 10.

Jencks, Christopher, et al. *Inequality*. New York: Basic Books, 1972.

Lee, Richard B., and Irven DeVore, eds. *Man the Hunter*. Chicago: Aldine, 1968.

Le Gros Clark, W. E. *History of the Primates*. Chicago: University of Chicago Press, 1959.

Mead, George Herbert. *The Philosophy of the Act*. Chicago: University of Chicago Press, 1938.

Mill, John Stuart. *Mill's Utilitarianism*. Edited by James M. Smith and Ernest Sosa. Belmont, Calif.: Wadsworth Publishing Co., 1969.

Modgil, Sohan, and Celia Modgil, eds. *Arthur Jensen: Consensus and Controversy*. New York: Falmer Press, 1987.

Ortega y Gasset, José. *The Revolt of the Masses*. London: Unwin Books, 1963.

Pelczynski, Zbigniew, and John Gray, eds. *Conceptions of Liberty in Political Philosophy*. London: Athlone Press, 1984.

Piaget, Jean. *Play, Dreams, and Imitation in Childhood*. London: Routledge and Kegan Paul, 1951.

Scheler, Max. *Ressentiment*, translated by W. W. Holdheim and edited by L. A. Coser. Glencoe, Ill.: Free Press, 1961.

Sherif, M., O. J. Harvey, B. J. White, W. R. Hood, and C. W. Sherif. *Intergroup Conflict and Cooperation: The Robbers Cave Experiment*. Norman, Okla.: Institute of Group Relations, University of Oklahoma, 1961.

Smith, E. O., ed. *Social Play in Primates*. New York: Academic Press, 1978.

Smith, P. K., ed. *Play in Animals and Humans*. Oxford: Basil Blackwell, 1984.

Somit, Albert, ed. *Biology and Politics*. Paris: Mouton, 1976.

Spengler, Oswald. *The Decline of the West*. Translated by Charlers Francis Atkinson. 2 vols. New York: Alfred A. Knopf, 1926, 1928.

Tobias, Phillip V., ed. *Hominid Evolution: Past, Present and Future*. New York: Alan R. Liss, 1985.

Tocqueville, Alexis de. *Democracy in America*. Translated by George Lawrence and edited by J. P. Mayer. New York: Harper and Row, 1988.

Watson, George. *The Idea of Liberalism: Studies for a New Map of Politics*. London: Macmillan, 1985.

Wilson, Edward O. *On Human Nature*. Cambridge, Mass.: Harvard University Press, 1978.

INDEX

ABOUT THE AUTHOR

MICHIO KITAHARA is currently the director of the Nordenfeldt Institute, in Gothenburg, Sweden, where his research focuses on individualism, liberalism, and freedom. He has held teaching or research appointments at the Universities of Maryland, Michigan, and San Francisco, as well as at the State University of New York at Buffalo. Dr. Kitahara, who was born in Japan, is the author of *Children of the Sun: The Japanese and the Outside World, Twelve Propositions on the Self: A Study of Cognitive Consistency in the Sociological Perspective, An Axiomatic Theory of Balance: A Study of Self in the Sociocultural Environment*, and *An Essay on Culture: A Definition of Culture and Its Implications to the Study of Sociocultural Dynamics*.